BIRKHÄUSER

Progress in Mathematics
Volume 268

Detlef Gromoll
Gerard Walschap

Metric Foliations and Curvature

Birkhäuser
Basel · Boston · Berlin

Authors:

Detlef Gromoll †
State University of New York
Stony Brook, USA

Gerard Walschap
Department of Mathematics
University of Oklahoma
Norman, OK 73019-0315
USA
e-mail: gerard@ou.edu

2000 Mathematics Subject Classification 53C12, 57R30

Library of Congress Control Number: 2008935487

Bibliographic information published by Die Deutsche Bibliothek.
Die Deutsche Bibliothek lists this publication in the Deutsche Nationalbibliografie; detailed bibliographic data is available in the Internet at http://dnb.ddb.de

ISBN 978-3-7643-8714-3 Birkhäuser Verlag AG, Basel · Boston · Berlin

Basel · Boston · Berlin
P.O. Box 133, CH-4010 Basel, Switzerland
Part of Springer Science+Business Media
Printed on acid-free paper produced from chlorine-free pulp. TCF ∞
Printed in Germany

ISBN 978-3-7643-8714-3 e-ISBN 978-3-7643-8715-0

9 8 7 6 5 4 3 2 1 www.birkhauser.ch

Contents

Preface

The theory of foliations is closely related to that of differential equations: an oriented one-dimensional foliation is equivalent to a nowhere vanishing vector field X on a manifold M, and the integral curves of X are solutions of a system of ordinary differential equations. Higher-dimensional foliations correspond to systems of partial differential equations via the Frobenius theorem. When the solutions (or leaves) are locally equidistant, the foliation is said to be metric. If in addition, the space B of leaves is reasonably well-behaved, the map $M \longrightarrow B$ that sends a point of M to the leaf on which it lies is called a metric fibration or Riemannian submersion.

In the past three or four decades, there has been increasing realization that these foliations play a key role in understanding the structure of Riemannian manifolds, particularly those with positive or nonnegative sectional curvature. In fact, all known such spaces are constructed from only a representative handful by means of metric fibrations or deformations thereof. This is even more pronounced in positive curvature, where every such space is the image of a Riemannian submersion from a nonnegatively curved manifold. Further indication of the key role that submersions play in nonnegative curvature is Perelman's result that all noncompact spaces with curvature ≥ 0 are metric fibrations over compact ones.

This text is an attempt to document some of these constructions, many of which have only appeared in journal form. The emphasis here is less on the fibration itself and more on how to use it to either construct or understand a metric with curvature of fixed sign on a given space. The approach differs in this sense from previous ones in which a typical question would be to ask whether there exists a metric on the ambient space for which a given foliation has this or that property. The reader will in fact find that this work has little intersection with other books on the subject such as Molino's [91] or Tondeur's [124]. In particular, topics such as basic cohomology or Lie foliations are either omitted or only briefly mentioned. It is assumed that the reader has a working knowledge of differentiable manifolds and Riemannian metrics, such as that offered in an intermediate level course in Riemannian geometry.

The first chapter introduces the main concepts and tools that are used throughout, and relates the curvature of the ambient space with that of the base.

It further discusses the relation between the geodesics in both manifolds as well as the Jacobi fields along them, and ends with the description of Wilking's dual foliation.

Chapter 2 begins by studying how warping of the fibers affects the curvature of the ambient space. This is then used to introduce warped products, after which we discuss the main class of Riemannian submersions, namely those generated by isometric group actions. The rest of the chapter is mostly devoted to the construction of spaces of positive or nonnegative curvature by means of submersions. This includes fundamental examples such as Lie groups, as well as more elaborate ones, such as the Allof-Wallach and Eschenburg spaces.

Chapter 3 studies the structure of complete, noncompact manifolds with curvature ≥ 0, beginning with the Cheeger-Gromoll soul construction. We present Perelman's proof of the generalized soul theorem, and Wilking's work on smoothness of the metric projection onto the soul. The converse of the soul theorem, namely the question of which vector bundles admit nonnegatively curved metrics is also discussed.

The last chapter deals with the problem of classifying metric foliations on spaces of constant curvature. Although this is a fundamental question, it is also a surprisingly delicate one which at the time of writing is still not entirely answered.

We would like to thank Taechang Byun and Luis Guijarro for reading preliminary versions of the manuscript and offering valuable suggestions and corrections.

On a tragic note, the first named author, Detlef Gromoll, passed away during the final revision phase of this manuscript. He is fondly remembered by his many friends, colleagues, and former students.

Chapter 1

Submersions, Foliations, and Metrics

The concept of submersion is dual to what is arguably the oldest notion in differential geometry, that of immersion. Both are generalizations of diffeomorphisms. In the presence of a Riemannian metric, it is natural to consider distance-preserving maps rather than diffeomorphisms. These in turn generalize to isometric immersions, and their metric dual, Riemannian submersions.

1.1 Notation and basic geometric concepts

In order to fix notation, we begin by briefly recalling some of the basic concepts that will be used throughout. For further details, the reader is referred for example to [104] or [136]. All maps and manifolds are assumed to be sufficiently smooth. The tangent space of a manifold M at $p \in M$ will be denoted by T_pM or often M_p, the algebra of real-valued functions $\phi : M \to \mathbb{R}$ on M by $\mathcal{F}(M) = \mathcal{F}_M = C^{\cdot}(M)$, and the Lie algebra of vector fields on M by $\mathfrak{X}(M) = \mathfrak{X}_M = \Gamma TM$; the last term in the previous identity refers to the space of sections of the tangent bundle TM of M. The derivative of a map $f : M \to N$ at p is the vector bundle homomorphism $f_* : TM \to TN$ from the tangent bundle TM of M into that of N which restricts fiber-wise to the linear map $f_{*p} : M_p \to N_{f(p)}$ given by $f_{*p}v(\phi) = v(\phi \circ f)$ for $p \in M$, $v \in M_p$, $\phi \in \mathcal{F}(N)$.

Let $I \subset \mathbb{R}$ be an interval, $D \in \mathfrak{X}(I)$ the standard coordinate vector field on I corresponding to the identity chart $(I, 1_I)$ of I. The *tangent field* of c is the vector field $\dot{c}$ along c defined by $\dot{c} = c_* D$.

Definition 1.1.1. Let M^{n+k} and B^n denote manifolds of dimension $n + k$ and n respectively. A surjective map $\pi : M \to B$ is said to be a *submersion* if its derivative π_{*p} at any $p \in M$ has maximal rank n.

By the implicit function theorem, the preimage $F := \pi^{-1}(b)$ of a point $b \in B$ is a k-dimensional submanifold of M, called the *fiber of* π *over* b, even though π need not be a fibration in the usual sense. If $\imath : F \to M$ denotes inclusion, and $p \in F$, then $\imath_* F_p = \ker \pi_{*p}$. For the sake of brevity, we will identify F_p with the subspace $\imath_* F_p$ of M_p. A fundamental example of a submersion is the tangent bundle projection $\pi_M : TM \to M$, where the fiber over $p \in M$ is the tangent space M_p of M at p.

Subbundles of the tangent bundle – such as the kernel of the derivative of a submersion $\pi : M \to B$ – of a manifold M are called *distributions* on M. A (local) *section* of a distribution $\mathcal{D}$ is a map $X : U \to \mathcal{D}$ from an open set U of M to the distribution such that $\pi_M \circ X = 1_U$, where 1_U denotes the identity map on U. A distribution $\mathcal{D}$ is said to be *integrable* if the space $\mathfrak{X}\mathcal{D}$ of sections of $\mathcal{D}$ is a Lie algebra under the usual Lie bracket of vector fields.

The Frobenius theorem asserts that if $\mathcal{D}$ is an integrable distribution on M, then each point p of M is contained in some *integral manifold* of $\mathcal{D}$; i.e., p belongs to an immersed submanifold N of M, the tangent bundle of which coincides with the restriction $\mathcal{D}|_N$ of the distribution to N. In fact, if $\mathcal{D}$ is k-dimensional, then there exists a chart (U, x) of M around p, with $x(U) = (-1,1)^{n+k}$, $x(p) = 0$, such that the slices $x^{-1}(0,a)$, for $(0,a) \in \{0\} \times (-1,1)^n \subset (-1,1)^{n+k}$, coincide with the maximal connected integral manifolds of $\mathcal{D}$ contained in U; in other words, these manifolds are given by the fibers of the submersion $\pi \circ x : U \to (-1,1)^n$, where $\pi : (-1,1)^k \times (-1,1)^n \to (-1,1)^n$ denotes projection.

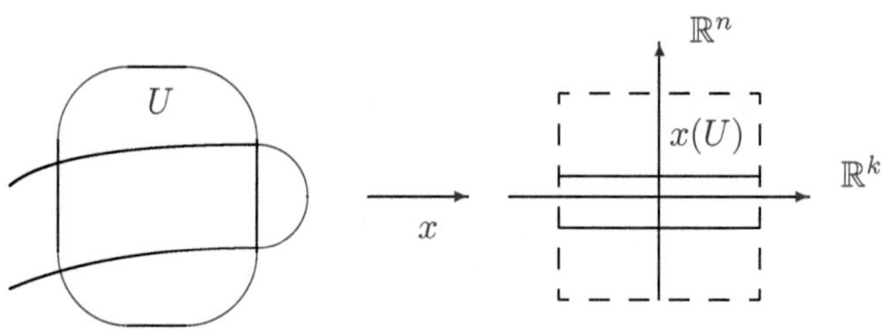

The Frobenius theorem can be conveniently reformulated in terms of differential forms: a differential l-form α is said to *annihilate* a k-dimensional distribution $\mathcal{D}$ if for any $p \in M$,

$$\alpha(p)(v_1, \dots, v_l) = 0 \quad \text{whenever } v_1, \dots, v_l \in \mathcal{D}_p.$$

Consider the ideal $I(\mathcal{D})$ in the exterior algebra of M of all forms that annihilate $\mathcal{D}$. Then $\mathcal{D}$ is integrable iff $I(\mathcal{D})$ is a *differential ideal*; i.e., iff $d(I(\mathcal{D})) \subset I(\mathcal{D})$. This is an immediate consequence of the fact that $\mathcal{D}$ is locally generated by n independent 1-forms $\omega^1, \dots, \omega^n$ which annihilate $\mathcal{D}$, together with the identity

$$\omega^i[X,Y] = -d\omega^i(X,Y) + X\omega^i(Y) - Y\omega^i(X) = -d\omega^i(X,Y)$$

for sections X, Y of $\mathcal{D}$. In fact, if $\mathcal{D}$ is spanned on some open set U by linearly independent sections $X_{n+1}, \dots, X_{n+k}$, extend the latter to linearly independent sections $X_1, \dots, X_{n+k}$ on (some open subset of) U. If ω^i denote the dual one-forms, then $I(D)$ is generated by (the kernels of) $\omega^1, \dots, \omega^n$:

$$I(D) \cap TU = \{v \in TU \mid \omega_i(v) = 0, \quad i = 1, \dots, n\}.$$

Furthermore, setting $\Omega = \omega^1 \wedge \cdots \wedge \omega^n$, the integrability condition is equivalent to

$$d\Omega = \alpha \wedge \Omega \quad \text{for some 1-form } \alpha, \tag{1.1.1}$$

since each $d\omega^i = \sum_j \alpha_{ij} \wedge \omega_j$ for some 1-forms α_{ij} on U.

Definition 1.1.2. Let $\mathcal{D}$ denote an integrable distribution on M. The collection of integral manifolds of $\mathcal{D}$ is called a *foliation* of M. A maximal connected integral manifold of $\mathcal{D}$ is called a *leaf* of the foliation.

For example, the collection of fibers of a submersion $\pi : M \to B$ is the foliation of M induced by the distribution $\ker \pi_*$. Conversely, the leaves of a foliation of M are always locally given by the fibers of a submersion $\pi \circ x : U \to (-1, 1)^n$ as above. Unlike submersions, the leaves of a foliation need not be imbedded submanifolds of M, as some of the examples below show.

Examples and Remarks 1.1.1.

(i) Let $u^i : \mathbb{R}^n \to \mathbb{R}$, $u^i(a_1, \dots, a_n) = a_i$, denote the projection onto the ith factor, and D_i the coordinate vector field $\partial/\partial u^i$ on $\mathbb{R}^n$ corresponding to the chart $(\mathbb{R}^n, 1_{\mathbb{R}^n})$ of $\mathbb{R}^n$, $i = 1, \dots, n$. Given a real number a, the vector field $X = -u^2 D_1 + u^1 D_2 - au^4 D_3 + au^3 D_4$ on $\mathbb{R}^4$ is tangent to the unit sphere $M = S^3$, and thus induces a one-dimensional (necessarily) integrable distribution on M. Viewing M as the set of all pairs of complex numbers $(z_1, z_2) \in \mathbb{C}^2$ with $|z_1|^2 + |z_2|^2 = 1$, the one-parameter group Φ_t of diffeomorphisms corresponding to the flow of X is given by

$$\Phi_t(z_1, z_2) = (z_1 e^{it}, z_2 e^{iat}), \quad t \in \mathbb{R}, \quad (z_1, z_2) \in S^3,$$

and the leaf through a point p is the orbit $\{\Phi_t(p) \mid t \in \mathbb{R}\}$ of the point. When $a = 1$, all the leaves are great circles, the space of leaves is $\mathbb{CP}^1 = S^2$, and the map $\pi : M \to S^2$ that assigns to a point the leaf it belongs to is a submersion, called the *Hopf fibration*. When a is, say, irrational, only the leaves through $\pm(1, 0)$ and $\pm(0, 1)$ are great circles; all others are immersed copies of $\mathbb{R}$.

(ii) Consider the torus $S^1_{1/\sqrt{2}} \times S^1_{1/\sqrt{2}} = \{(z_1, z_2) \in S^3 \mid |z_1|^2 = |z_2|^2 = 1/2\}$. The foliation in (i) restricts to a foliation on the torus. It is easy to see that all the leaves are dense if a is irrational.

(iii) The torus T in (ii) partitions the 3-sphere into two disjoint open sets $U_1 = \{(z_1, z_2) \in S^3 \mid |z_1|^2 < 1/2\}$ and $U_2 = \{(z_1, z_2) \mid |z_2|^2 < 1/2\}$ with common boundary T. These two sets are diffeomorphic via $(z_1, z_2) \mapsto (z_2, z_1)$, and they have as closure a solid torus $D^2 \times S^1$, where D^2 denotes the unit disk in the plane. In fact, the map $(z_1, z_2) \mapsto (\sqrt{2} z_1, z_2/|z_2|)$ is a diffeomorphism from $\bar{U}_1$ onto $D^2 \times S^1$. Thus, any two-dimensional foliation of the solid torus that has the boundary T as a leaf induces a two-dimensional foliation of the 3-sphere. We describe one such foliation below. The resulting foliation of S^3 is known as the *Reeb foliation*.

Consider cylindrical coordinates (r, θ, z) on $D^2 \times \mathbb{R}$. Let ϕ be a smooth function on $\mathbb{R}$ that is 0 when $r \leq 0$, 1 when $r \geq 1$, and such that $\phi(r) \in (0,1)$ when $r \in (0,1)$. If ω is the 1-form on $D^2 \times \mathbb{R}$,

$$\omega = \phi(r)dr + (1 - \phi(r))dz,$$

then

$$d\omega = -\phi'(r)dr \wedge dz = \alpha \wedge \omega,$$

where $\alpha = (2(1-\phi) + \phi')dz + (2\phi - \phi')dr$. By (1.1.1), the two-dimensional distribution corresponding to the kernel of ω is integrable, and the resulting foliation on $D^2 \times \mathbb{R}$ induces one on the solid torus $D^2 \times S^1$. Since the boundary T of the torus is a leaf, we obtain a two-dimensional foliation of S^3 with exactly one compact leaf.

(iv) Let G be a Lie group, H a closed subgroup of G. There exists a unique differentiable structure on the quotient space G/H for which the natural projection $\pi : G \to G/H$ becomes a submersion. G/H is then called a *homogeneous space*.

(v) Let $\xi = \pi : E \to M$ be a vector bundle over M. A (linear) *connection* $\mathcal{H}$ on ξ is a distribution on the total space E such that if $\mathcal{H}_u$ denotes the fiber of the subbundle $\mathcal{H}$ over $u \in E$, then

1. $\pi_{*u} : \mathcal{H}_u \to M_{\pi(u)}$ is an isomorphism for all $u \in E$, and
2. $\mu_{a_*}\mathcal{H}_u = \mathcal{H}_{au}$, where $\mu_a(u) = au$ is multiplication by $a \in \mathbb{R}$.

The first condition above implies that the tangent bundle of E decomposes as a direct sum $TE = \mathcal{H} \oplus \ker \pi_*$. The connection is said to be *flat* if it is integrable. This is a fairly restrictive condition, since it implies for example that E admits two complementary foliations.

(vi) A surjective map $\pi : M \to B$ is said to be a *fibration* if it has the *homotopy lifting property*: namely, given a manifold N, and a map $f : N \to M$, any homotopy $H : N \times [0,1] \to B$ of $\pi \circ f$ can be lifted to a homotopy $\tilde{H} : N \times [0,1] \to M$ of f; i.e., $\pi \circ \tilde{H} = H$, and $\tilde{H} \circ \imath_0 = f$, where $\imath_t : N \to N \times [0,1]$ maps p to (p,t), $0 \leq t \leq 1$. In other words, one must be able to fill in the dashed arrow in the following diagram:

$$\begin{array}{ccc} N = N \times \{0\} & \xrightarrow{\ f\ } & M \\ {\scriptstyle \imath_0}\downarrow & \overset{\tilde{H}}{\nearrow} & \downarrow{\scriptstyle \pi} \\ N \times [0,1] & \xrightarrow[H]{} & B \end{array}$$

A fibration is necessarily a submersion: let p be in M, $b := \pi(p)$. It must be shown that π_{*p} is surjective. So let $v \in B_b$, $c : [0,1] \to B$ be a curve in B with $c(0) = b$, $\dot{c}(0) = v$. Since c is a homotopy of $\pi \circ f$, where $f : \{0\} \to M$ maps 0 to p, there exists a curve $\tilde{c}$ in M that starts at p and projects to c. Then $\pi_*(\dot{\tilde{c}}(0)) = v$, as claimed. On the other hand, the projection $u^1 : \mathbb{R}^2 \setminus \{0\} \to \mathbb{R}$ onto the x-axis is a submersion which is not a fibration. It can be shown that $\pi : M \to B$ is a fibration if it is one locally; i.e., if each point of B has a neighborhood U such that the restriction $\pi : \pi^{-1}(U) \to U$ is a fibration, cf. [26]. In particular, a locally trivial fiber bundle (a map $\pi : M \to B$ such that each point of B has a neighborhood U and a diffeomorphism $(\pi, \phi) : \pi^{-1}(U) \to U \times F$ for some fixed F) is easily seen to be a fibration. The converse, though, is in general not true, see [51], [85] for examples of fibrations that are not locally trivial fiber bundles.

1.2 Metric foliations and Riemannian submersions

The concept of Riemannian submersion is a special case of an elementary notion that can best be described in the context of metric spaces: Let (M, d) denote a metric space. A *singular metric foliation* $\mathcal{F}$ of M is a decomposition of M into connected subsets, called leaves, that are locally equidistant; i.e., for any $p \in M$, there exist neighborhoods $U \subset V$ of p such that the following holds: given two leaves L_i and connected components N_i of $L_i \cap V$, $i = 1, 2$, the distance function $q \mapsto d(q, N_1)$ is constant on $N_2 \cap U$. In the case when $U = V = M$, the orbit space $M/\mathcal{F}$ inherits a metric from d, and the projection $\pi : M \to M/\mathcal{F}$ is a *submetry*: that is, π maps any closed metric ball around $p \in M$ onto the metric ball of same radius around $\pi(p)$[22]. Notice that we do not require the existence of an isometry, or even a bijection, between two of these subsets. For example, consider the 2-sphere of radius r with its canonical metric, and decompose it into circles of latitude. The north and south poles can then be viewed as degenerate circles, and the quotient metric space is a closed interval of length πr.

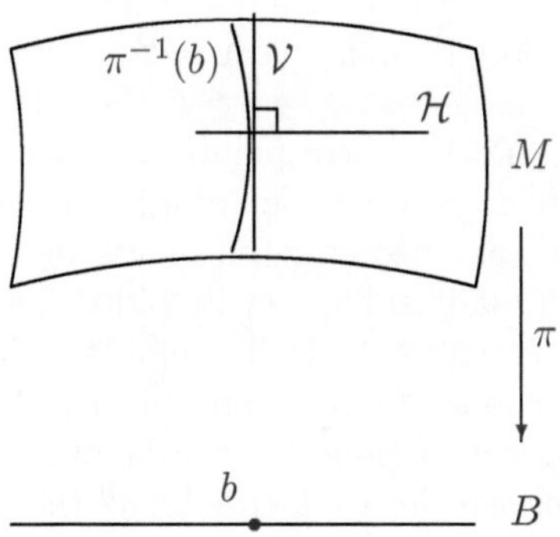

We now wish to examine these concepts in the context of Riemannian manifolds. The *vertical distribution* $\mathcal{V}$ of a submersion $\pi : M \to B$ is defined to be the kernel of π_*, i.e., the collection of tangent spaces to the fibers. A *vertical field* is a section of the vertical subbundle $\mathcal{V} \to M$, or more simply put, a vector field on M that is everywhere tangent to the leaves of the submersion. The space $\mathfrak{X}^{\mathbf{v}}$ of vertical fields on M is a Lie subalgebra of $\mathfrak{X}(M)$. It is in fact an ideal of the algebra of *projectable fields* on M; i.e., those vector fields that are π-related to vector fields on B: If $E \in \mathfrak{X}(M)$ is π-related to $X \in \mathfrak{X}(B)$ and U is vertical, then U is π-related to the trivial field on B, and $\pi_*[E,U] = [X,0] \circ \pi = 0$, so $[E,U]$ is vertical. Notice that any complementary subspace to $\ker \pi_{*p}$ in M_p is mapped isomorphically onto the tangent space of B at $\pi(p)$. In general, there is no canonical complementary subspace to $\mathcal{V}_p$. If M is a Riemannian manifold, however, one such is the orthogonal complement of $\mathcal{V}_p$.

Definition 1.2.1. Let $\pi : M \to B$ be a submersion, where M is a Riemannian manifold. The *horizontal distribution* of π is the orthogonal complement $\mathcal{H} = \mathcal{V}^{\perp}$ of $\mathcal{V}$. If in addition B is a Riemannian manifold, then the submersion is said to be *Riemannian* if it is isometric when restricted to the horizontal distribution; i.e., if $|\pi_* x| = |x|$ for all $x \in \mathcal{H}$.

Thus, Riemannian submersions generalize isometries to the case when $n \geq k$, just as isometric immersions generalize them for $n \leq k$. Riemannian manifolds are, of course, metric spaces, and it makes sense to consider submetries (as defined above) between them. It can be shown that a submetry between Riemannian manifolds is a $C^{1,1}$ Riemannian submersion, cf. [19].

Given a Riemannian submersion $\pi : M \to B$, the orthogonal splitting of the tangent bundle of M induces a decomposition $e = e^{\mathbf{h}} + e^{\mathbf{v}} \in \mathcal{H} \oplus \mathcal{V}$ of any $e \in TM$. We will for the most part abbreviate $(\nabla_E F)^{\mathbf{h}}$ by $\nabla^{\mathbf{h}}_E F$, and similarly for the vertical component. A *basic* vector field on M is one that is both horizontal and projectable. The space $\mathcal{B}$ of basic fields is isomorphic to $\mathfrak{X}(B)$, but is not, in general, a Lie algebra. Notice though, that since elements of $\mathcal{B}$ are projectable, $[\mathcal{B}, \mathfrak{X}^{\mathbf{v}}] \subset \mathfrak{X}^{\mathbf{v}}$ as above; i.e.,

$$[X,U]^{\mathbf{h}} = 0, \qquad X \in \mathcal{B}, \quad U \in \mathfrak{X}^{\mathbf{v}}. \tag{1.2.1}$$

One would like to have a corresponding notion for foliations. In this case, there is no base manifold, however, let alone a Riemannian one. Nevertheless, according to the Frobenius theorem, the leaves of a foliation are locally given by fibers of submersions, so it suffices to consider the following question: given a submersion $\pi : M \to B$, where M is Riemannian, does there exist a metric on the base for which π becomes Riemannian? Intuitively, a necessary and sufficient condition should be that the horizontal metric is invariant under the flow of vertical fields.To render this precisely, let us denote by $g^{\mathbf{h}}$ the horizontal component of the metric tensor: $g^{\mathbf{h}}(E,F) = \langle E^{\mathbf{h}}, F^{\mathbf{h}} \rangle$, E, $F \in \mathfrak{X}(M)$. Horizontal vector fields will be denoted by X, Y, Z, vertical ones by U, V, W.

Theorem 1.2.1. *Let $\pi : M \to B$ be a submersion with connected fibers, where M is a Riemannian manifold. Then there exists a metric on B for which π becomes Riemannian iff the Lie derivative $\mathcal{L}_U g^{\mathbf{h}}$ of $g^{\mathbf{h}}$ vanishes in any vertical direction U.*

Proof. Suppose the Lie derivative $\mathcal{L}_U g^{\mathbf{h}}$ is zero. Then

$$0 = (\mathcal{L}_U g^{\mathbf{h}})(X,Y) = U\langle X,Y\rangle - \langle [U,X]^{\mathbf{h}},Y\rangle - \langle X,[U,Y]^{\mathbf{h}}\rangle. \tag{1.2.2}$$

Now, let $\bar{X}$, $\bar{Y}$ denote (local) vector fields on B, and consider their basic lifts X, Y to M; i.e., for $p \in M$, X_p is the unique horizontal vector that projects to $\bar{X}_{\pi(p)}$ via π_* (smoothness of basic lifts will be established in the next section). Then $[U,X]^{\mathbf{h}} = [U,Y]^{\mathbf{h}} = 0$, and by (1.2.2), $U\langle X,Y\rangle = 0$ for any $U \in \mathfrak{X}^{\mathbf{v}}$, so that $\langle X,Y\rangle$ is constant along fibers. We may therefore define the metric on B by $\langle \bar{X},\bar{Y}\rangle := \langle X,Y\rangle$. Conversely, suppose $\pi : M \to B$ is a Riemannian submersion, and X, $Y \in \mathcal{B}$. By (1.2.1) and the definition of the Levi-Civita connection,

$$\begin{aligned} 2\langle \nabla^{\mathbf{v}}_X Y, U\rangle &= X\langle Y,U\rangle + Y\langle U,X\rangle - U\langle X,Y\rangle \\ &\quad + \langle U,[X,Y]\rangle + \langle X,[Y,U]\rangle - \langle Y,[U,X]\rangle \\ &= \langle [X,Y]^{\mathbf{v}},U\rangle, \end{aligned}$$

and $\nabla^{\mathbf{v}}_X Y = (1/2)[X,Y]^{\mathbf{v}}$ for basic X, Y. Now, the map $[,]^{\mathbf{v}} : \mathfrak{X}^{\mathbf{h}} \times \mathfrak{X}^{\mathbf{h}} \to \mathfrak{X}^{\mathbf{v}}$ is $\mathcal{F}(M)$-linear, since for $\phi \in \mathcal{F}(M)$, $[\phi X,Y]^{\mathbf{v}} = \phi[X,Y]^{\mathbf{v}} - (Y\phi)X^{\mathbf{v}} = \phi[X,Y]^{\mathbf{v}}$. Similarly, $\nabla^{\mathbf{v}} : \mathfrak{X}^{\mathbf{h}} \times \mathfrak{X}^{\mathbf{h}} \to \mathfrak{X}^{\mathbf{v}}$ is $\mathcal{F}(M)$-linear, so that

$$\nabla^{\mathbf{v}}_X Y = \frac{1}{2}[X,Y]^{\mathbf{v}}, \qquad X,Y \in \mathfrak{X}^{\mathbf{h}}, \tag{1.2.3}$$

and in particular, the operator $\nabla^{\mathbf{v}} : \mathcal{H} \times \mathcal{H} \to \mathcal{V}$ is a skew-symmetric tensor field, a fact that will be important in the sequel. This in turn implies that for horizontal X, Y, and vertical U,

$$\begin{aligned} (\mathcal{L}_U g^{\mathbf{h}})(X,Y) &= U\langle X,Y\rangle - \langle [U,X],Y\rangle - \langle X,[U,Y]\rangle \\ &= \langle \nabla_X U,Y\rangle + \langle \nabla_Y U,X\rangle = -\langle U, \nabla^{\mathbf{v}}_X Y + \nabla^{\mathbf{v}}_Y X\rangle \\ &= 0 \end{aligned}$$

by (1.2.3). □

Implicit in the proof of Theorem 1.2.1 is the following:

Remark 1.2.1. Let $\mathcal{F}$ be a foliation on a Riemannian manifold. Then $\mathcal{L}_U g^{\mathbf{h}}$ vanishes horizontally for all $U \in \mathfrak{X}^{\mathbf{v}}$ iff $\nabla^{\mathbf{v}}_X X = 0$ for all $X \in \mathfrak{X}^{\mathbf{h}}$.

In particular, the leaves of such a foliation are locally given by fibers of Riemannian submersions for adequately defined metrics on the local quotients.

Definition 1.2.2. A foliation on a Riemannian manifold is said to be *metric* if $\mathcal{L}_U g^{\mathbf{h}}$ is horizontally zero for any $U \in \mathfrak{X}^{\mathbf{v}}$, or equivalently, if $\nabla^{\mathbf{v}} : \mathcal{H} \times \mathcal{H} \to \mathcal{V}$ is skew-symmetric.

Examples and Remarks 1.2.1. (i) We have used the terminology "metric foliation" in Definition 1.2.2 instead of the more traditional "Riemannian foliation" because the latter is sometimes reserved for foliations on manifolds that are not necessarily endowed with a global Riemannian metric – i.e., a Euclidean metric on TM – but rather one on the quotient bundle $TM/\mathcal{V}$ only, cf. for example [124]. This transversal metric is then required to satisfy the same condition that g^{h} does in our case. In this book, however, a Riemannian foliation always refers to a metric foliation.

(ii) Let $\mathcal{F}$ be a foliation of dimension k on M^{n+k}, $\{(U_\alpha, x_\alpha)\}_{\alpha\in A}$ an atlas of charts on M such that for $x_\alpha : U_\alpha \to (-1,1)^{n+k}$, the connected components of the leaves of $\mathcal{F}$ in U_α are given by the slices $x_\alpha^{-1}(0,a)$, $(0,a) \in \{0\}\times(-1,1)^n \subset \mathbb{R}^{n+k}$. Denote by π_2 the projection $\mathbb{R}^{n+k} \to \mathbb{R}^n$, and let $\pi_\alpha = \pi_2 \circ x_\alpha : U_\alpha \to (-1,1)^n$, so that the slices coincide with the fibers of the submersion π_α. For any α, β in A, there is a diffeomorphism $f_{\alpha\beta} : \pi_\alpha(U_\alpha \cap U_\beta) \to \pi_\beta(U_\alpha \cap U_\beta)$ such that

$$f_{\alpha\beta} \circ \pi_\alpha = \pi_\beta. \tag{1.2.4}$$

If M is Riemannian and $\mathcal{F}$ is a metric foliation, then by Theorem 1.2.1, there exists a Riemannian metric on each $\pi_\alpha(U_\alpha)$ for which π_α becomes a Riemannian submersion. (1.2.4) then implies that each transition function $f_{\alpha\beta}$ is isometric.

(iii) Suppose U is a Killing vector field on M with no zeros. Then $\mathcal{L}_U g = 0$, and the one-dimensional foliation generated by the integral curves of U is metric. The foliation in Examples and Remarks 1.1.1(i) is a metric foliation of this type, since the one-parameter group of the vector field consists of isometries of S^3.

(iv) More generally, let G be a subgroup of the isometry group of a Riemannian manifold M, and suppose that all orbits have the same type (meaning that any two are equivariantly diffeomorphic), so that there exists a differentiable structure on the space M/G of orbits for which the natural projection $\pi : M \to M/G$ is a submersion. For $g \in G$, denote by $L_g : M \to M$ the isometry mapping p to $g(p)$, and for $p \in M$, denote by $\lambda_p : G \to M$ the map that sends g to $g(p)$; i.e., λ_p maps G onto the orbit $G(p) = \pi^{-1}(\pi(p))$ of p. Then each element U in the Lie algebra $\mathfrak{g}$ of G induces a vertical vector field $\tilde{U}$ on M defined by

$$\tilde{U}(p) = \lambda_{p*e} U(e), \qquad p \in M,$$

and the collection of these vector fields span the fibers of π. We claim that each such field is a Killing field, so that there exists a unique metric on M/G for which π becomes Riemannian: To see this, let $\phi : \mathbb{R} \to G$ denote the Lie group homomorphism with $\dot{\phi}(0) = U(e)$, and for $q \in M$, consider the curve $t \mapsto c_q(t) := L_{\phi(t)}(q)$. For any fixed $t_0 \in \mathbb{R}$, we have

$$\begin{aligned} c_q(t) = L_{\phi(t)}(q) = \lambda_q(\phi(t)) = \lambda_q(\phi(t-t_0)\phi(t_0)) = \lambda_{\phi(t_0)q}(\phi(t-t_0)) \\ = \lambda_{c_q(t_0)}(\phi(t-t_0)). \end{aligned}$$

Thus, $\dot{c}_q(t_0) = \lambda_{c_q(t_0)*e}\dot{\phi}(0) = \lambda_{c_q(t_0)*e}U(e) = \tilde{U} \circ c_q(t_0)$, and c_q is the integral curve of $\tilde{U}$ passing through q at $t = 0$. In other words, the flow of $\tilde{U}$ is the one-parameter group of isometries $L_{\phi(t)}$, and $\tilde{U}$ is Killing.

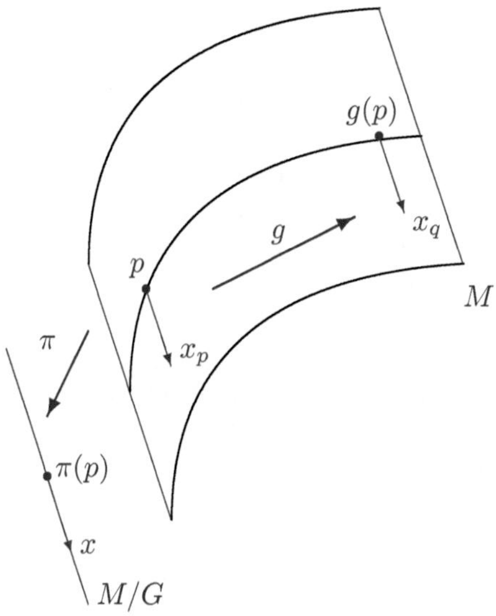

We could actually have argued this more directly by explicitly exhibiting the metric on M/G: In order to define the length of a vector x in the tangent space of M/G at some point $\pi(p)$, consider the orthogonal complement $\mathcal{H}_p$ of the tangent space to the fiber $\pi^{-1}(\pi(p))$ at p. This subspace contains a unique x_p such that $\pi_* x_p = x$. Define $|x| := |x_p|$. To see that this is independent of the point p in the preimage, let q be any other point in the orbit $G(p)$ of p, and x_q the vector in $\mathcal{H}_q$ that gets mapped to x via π_*. By assumption, there exists some $g \in G$ mapping p to q. g preserves orbits, hence also their orthogonal complement, so that $g_*\mathcal{H}_p = \mathcal{H}_q$. On the other hand, $\pi \circ g = \pi$, which implies that $\pi_* g_* x_p = \pi_* x_p = x = \pi_* x_q$. But both $g_* x_p$ and x_q belong to $\mathcal{H}_q$, and the restriction of π is one-to-one on this subspace, so that the two vectors coincide. Since g is an isometry, it follows that $|x_p| = |x_q|$, and the norm of x is well defined. On the more elementary metric space level, notice also that any two orbits in M are necessarily equidistant, so that the orbit space immediately inherits a metric space structure from that of M.

(v) We have already remarked that the leaves of a foliation $\mathcal{F}$ need not share the same topology. In the case of a metric foliation on a compact manifold M, though, Reinhart has shown that the leaves have the same universal cover [109]. In this case, Molino [89] has shown that in a neighborhood of any leaf that is not closed, there exist *transverse Killing fields* (i.e., basic fields that project to Killing fields in a local quotient) the flows of which fill out the closure of the leaf. This result in turn uses a general construction that consists in lifting the foliation to a principal bundle over M. More precisely, define the *bundle of horizontal orthonormal frames* $P(M, \mathcal{F})$ over M to be the principal $O(n)$-bundle (here, n is the codimension of $\mathcal{F}$) associated to the vector bundle $\mathcal{H}$ over M. The lifted foliation on $P(M, \mathcal{F})$ can be described as follows: given $p \in M$, and a local submersion $\pi : U \longrightarrow B$ defining $\mathcal{F}$ in a neighborhood of p, there is a natural submersion $\rho : P(U, \mathcal{F}) \longrightarrow P(B)$ onto the bundle $P(B)$ of orthonormal frames of B, that takes the frame $x_1, \dots, x_n$ to $\pi_* x_1, \dots, \pi_* x_n$, see [91]. The leaves of the foliation are then locally defined by the submersion ρ; i.e., they have the form $\rho^{-1}(q)$, $q \in P(B)$. Returning to the original foliation on a compact simply connected M, Ghys has shown that there is always at least one compact leaf, provided the Euler characteristic of M is nonzero [52]. If none of the leaves are compact, then $\mathcal{F}$ is contained inside a higher-dimensional metric foliation $\mathcal{F}'$ that has at least one compact leaf, and that has the same leaf closures as the original $\mathcal{F}$. In fact, $\mathcal{F}'$ is obtained by considering the flows of an abelian Lie algebra of global transverse Killing fields [93].

(vi) If N is a codimension one submanifold of a Riemannian manifold M, then at least locally, N is a leaf of metric foliation: given $p \in N$, there exists a parallel section X of the normal bundle of N on a neighborhood U of p in N, and $\epsilon > 0$ such that the sets $L_s := \{\exp(sX(q)) \mid q \in U\}$, $|s| < \epsilon$, are the leaves of a codimension one metric foliation on a neighborhood of p, cf. also Examples 2.2.1. When M is complete and has positive curvature (or more generally nonnegative curvature everywhere and positive curvature at one point), we will later see that the foliation cannot be extended to all of M. Without curvature assumptions, metric foliations of codimension greater than one generically do not exist even locally, in the sense that any metric that admits one can be residually perturbed to metrics that do not.

1.3 Horizontal lifts and transversal holonomy

Let $\pi : M^n \to B^k$ denote a submersion. We begin by investigating the problem of lifting curves in B to M. At this stage, no Riemannian metric is required. Recall that a *lift* of $c : I \to B$ to M is a curve $\bar{c} : I \to M$ with $\pi \circ \bar{c} = c$. Lifts are in general not unique, and we therefore only consider those that have all their tangent vectors in a given distribution $\mathcal{H}$ complementary to the kernel of π_*. Such a curve will be called an *$\mathcal{H}$-lift* or a *horizontal lift.* The splitting $TM = \mathcal{H} \oplus \ker \pi_*$ induces smooth projections $\mathsf{p}^{\mathbf{h}}$ and $\mathsf{p}^{\mathbf{v}}$ onto the subbundles $\mathcal{H}$ and $\ker \pi_*$.

It may be assumed without loss of generality that c is a *regular* curve; i.e., that $\dot{c}(t) \neq 0$ for all $t \in I$: for otherwise, its "graph" $c_1 : I \to I \times B$, $c_1(t) = (t, c(t))$ is a regular curve, and a $(TI \oplus \mathcal{H})$-lift of c_1 for the submersion $(1_I, \pi) : I \times M \to I \times B$ has the form $t \mapsto (t, \bar{c}(t))$, where $\bar{c}$ is the desired lift of c. We first establish smoothness of basic lifts of vector fields.

Lemma 1.3.1. *Let $\pi : M \to B$ be a submersion, $\mathcal{H}$ a distribution complementary to* $\ker \pi_*$*, and X a vector field on B. Then the basic lift of X is smooth.*

Proof. Since the statement is local, we need only establish it near any point p of M. By the implicit function theorem, we may assume, up to a local diffeomorphism, that there exists a neighborhood of p of the form $V = U \times N$ on which the restriction of π is the projection $U \times N \to U$ onto the first factor. Then $\mathsf{p}^{\mathbf{h}}(X_{|U}, 0)$ is a smooth vector field on V which by construction is the basic lift of $X_{|U}$. □

Proposition 1.3.1. *Let $\pi : M \to B$ be a submersion, $\mathcal{H}$ a distribution complementary to* $\ker \pi_*$, *and $c : I \to B$ a curve. Then for any $t_0 \in I$ and any $p \in \pi^{-1}(c(t_0))$, there exists $\epsilon > 0$ and a horizontal lift $\bar{c} : [t_0, t_0 + \epsilon) \to M$ of $c_{|[t_0,t_0+\epsilon)}$ with $\bar{c}(t_0) = p$. Any two such lifts coincide on the intersection of their domains. If, furthermore, $I = [a, b]$ and M is compact (or M is a complete Riemannian manifold), then $\bar{c}$ is defined on all of $[a, b]$.*

Proof. As noted earlier, we may assume that c is a regular curve. Choose a vector field X on B such that $X \circ c = \dot{c}$ on some neighborhood of t_0, and consider the basic lift $\bar{X}$ of X to M. Integral curves $\bar{c}$ of $\bar{X}$ are by definition horizontal, and furthermore project down to integral curves of X, since

$$(\dot{\pi \circ \bar{c}}) = \pi_* \dot{\bar{c}} = \pi_* \bar{X} \circ c = X \circ \pi \circ \bar{c}.$$

Now (the restriction of) c itself is an integral curve of X, and the first claim follows. The others are immediate consequences. □

Returning to the Riemannian case, we can now establish the following facts that will be used many times throughout the sequel:

Theorem 1.3.1. *Let $\pi : M \to B$ denote a Riemannian submersion. If $c : I \to M$ is a geodesic with $\dot{c}(t_0) \in \mathcal{H}$ for some $t_0 \in I$, then $\dot{c}(t) \in \mathcal{H}$ for all $t \in I$, and $\pi \circ c$ is a geodesic in B. Such a c will be called a* horizontal geodesic *of M. Furthermore, if M is complete, then*

1. *B is complete;*
2. *π is a* submetry; *i.e., π maps the closure of the metric ball $B_r(p) = \{q \in M \mid d(p, q) < r\}$ of radius r around p onto the closure of $B_r(\pi(p))$ for any $p \in M$;*
3. *the fibers of π are equidistant; i.e., for any two fibers F_0 and F_1, and $p \in F_0$, the distance between p and F_1 equals that between F_0 and F_1;*
4. *π is a locally trivial fiber bundle; i.e., any point b in B has a neighborhood U such that $\pi^{-1}(U)$ is diffeomorphic to $U \times F$, where $F = \pi^{-1}(b)$.*

Proof. Suppose $c : I \to M$ is a geodesic, with $t_0 \in I$, $p := c(t_0)$, and $x := \dot{c}(t_0) \in \mathcal{H}$. Choose some interval $J \subset I$ around t_0 on which the geodesic c_B in B with $\dot{c}_B(t_0) = \pi_* x$ is defined and minimal. By Proposition 1.3.1, c_B admits a horizontal lift c_M in M with $c_M(t_0) = p$ on some subinterval J' of J. We will show that c_M is length-minimizing, and hence a geodesic. By uniqueness of geodesics, c_M must then coincide with $c|_{J'}$, and in particular will be horizontal. To see that c_M is indeed length-minimizing, suppose that $c_0 : [a, b] \to M$ is some other curve in M with the same endpoints as c_M. If L denotes the length function, then

$$L(c_0) = \int_a^b |\dot{c}_0| \geq \int_a^b |\dot{c}_0^{\mathbf{h}}| = \int_a^b |\pi_* \dot{c}_0| = L(\pi \circ c_0) \geq L(c_B) = L(c_M),$$

where the last inequality follows from the fact that c_B is minimal.

Suppose next that M is complete. Statements (1) and (2) are then an immediate consequence of the above. In order to prove (3), let $b_i := \pi(F_i)$, $i = 0, 1$; by (1), there exists a minimal normal (that is, unit-speed) geodesic $c : [0, a] \to B$ from b_0 to b_1. Now, the distance between any point of F_0 and any point of F_1 is at least as large as the length of c because π is distance-decreasing. Consequently,

$$d(F_0, F_1) \geq L(c), \tag{1.3.1}$$

where d denotes the distance function. On the other hand, for any $p \in F_0$, the horizontal lift of c starting at p is a curve which ends at some point of F_1 and has the same length as c. Thus,

$$d(F_0, F_1) \leq d(p, F_1) \leq L(c). \tag{1.3.2}$$

The claim now follows by comparing equations (1.3.1) and (1.3.2).

For the last statement, we will show that given $b \in B$, there exists a neighborhood U of b in B such that $\pi^{-1}(U)$ is diffeomorphic to $U \times F$, where $F = \pi^{-1}(b)$, cf. Examples and Remarks 1.1.1(v). Let $\epsilon > 0$ be the injectivity radius of B at b, and V the metric ball of radius ϵ centered at the origin in the tangent space B_b. For each $x \in V$, denote by X the section of the normal bundle of F in M with $\pi_* X = x$. Then the map

$$\begin{aligned} h : F \times V &\longrightarrow B_\epsilon(F) = \pi^{-1}(B_\epsilon(b)), \\ (p, x) &\longmapsto \exp X(p) \end{aligned}$$

is well defined by (1) and differentiable. Given $q \in B_\epsilon(F)$, consider the unique normal minimal geodesic $c : [0, a] \to B$ from $\pi(q)$ to b. If c_M denotes the horizontal lift of c starting at q, then $c_M(a)$ is well defined by completeness of M, and $p := c_M(a) \in F$. Then $q = h(p, x)$, where $x = -\dot{c}(a)$, and h is surjective. By uniqueness of horizontal lifts, h is injective. Being defined in terms of the exponential map, this in turn implies that h has maximal rank. Thus, the composition

$$\pi^{-1}(B_\epsilon(b)) \xrightarrow{h^{-1}} F \times V \xrightarrow{1_F \times \exp_b} F \times B_\epsilon(b)$$

is a diffeomorphism. □

From now on, we will use the term fibration to refer to a locally trivial fiber bundle, even though a fibration in the traditional sense of Examples and Remarks 1.1.1(vi) is not always a locally trivial fiber bundle. A *metric fibration* is just a synonym of Riemannian submersion. Riemannian manifolds will routinely be assumed to be complete, unless otherwise specified.

Definition 1.3.1. Let $\pi : M \to B$ be a metric fibration, and $c : [0, 1] \to B$ a piecewise smooth curve in the base space. The *holonomy diffeomorphism* associated to c is the map $h_c : \pi^{-1}(c(0)) \to \pi^{-1}(c(1))$ between the fibers over the endpoints of c that maps a point p in the first fiber to the endpoint of the horizontal lift of c that starts at p.

Observe that the inverse of h_c is h_{-c}, where $-c(t) := c(1-t)$.

Definition 1.3.2. The holonomy group $\mathrm{Hol}(b)$ of a metric fibration $\pi : M \to B$ at a point b in B is the group of holonomy diffeomorphisms h_c of the fiber over b, where c is a piece-wise smooth closed curve at b.

If b_0 and b_1 are points in B and c is a curve joining b_0 to b_1, then the map

$$\begin{aligned} \mathrm{Hol}(b_0) &\longrightarrow \mathrm{Hol}(b_1), \\ h_{c_0} &\longmapsto h_c \circ h_{c_0} \circ h_c^{-1} \end{aligned}$$

is an isomorphism of holonomy groups.

Examples and Remarks 1.3.1. (i) Proposition 1.3.1 is no longer true in general if one removes the compactness assumption. The same can be said of Theorem 1.3.1 without the completeness hypothesis: If $\pi : M \to B$ is a submersion with $\dim M > 1$, $\dim B < \dim M$, and $p \in M$, then the restriction of π to $M \setminus \{p\}$ is still a submersion, but in the topological case, some lifts exist only locally, whereas in the Riemannian case, π is no longer a fibration.

(ii) Let $F = \pi^{-1}(b)$ denote a fiber of a Riemannian submersion $\pi : M \to B$, $b \in B$, $\nu(F)$ the normal bundle of F in M. Given x in the tangent space of B at b, consider the basic section X of $\nu(F)$ with $\pi_* X = x$. If $x_1, \dots, x_k$ is a basis of B_b, then $X_1, \dots, X_k$, when evaluated at $p \in F$, span the fiber of $\nu(F)$ at p, so that $\nu(F)$ is a trivial bundle. The *Bott connection* ∇^B on $\nu(F)$ is the unique connection for which basic fields are parallel. ∇^B is Riemannian, since $\langle X, Y\rangle$ is constant along F for basic X and Y. If $\Gamma(\nu(F))$ denotes the space of sections of $\nu(F)$, then the Bott connection is given by

$$\nabla^B_u X = [U, X]^{\mathbf{h}}, \qquad u \in TF, \quad X \in \Gamma(\nu(F)), \tag{1.3.3}$$

where U is any local vertical field extending u. To see this, consider a basis X_i of basic sections, and write $X = \sum \phi^i X_i$, $\phi_i \in \mathcal{F}(F)$. Since $[U, X_i]$ is vertical by (1.2.1), both sides of (1.3.3) yield the same expression, namely $\sum u(\phi^i) X_i$, thereby establishing the claim.

The Bott connection actually makes sense in the context of foliations, not just submersions, and is usually viewed as a connection on the vector bundle $\mathcal{H}$ over M. The Jacobi identity for brackets implies that the corresponding curvature tensor vanishes. This in turn yields the existence of certain characteristic classes of the vector bundle $\mathcal{H}$; i.e., cohomology classes of M that depend only on the foliation, see for example [83], [78].

(iii) In the case of a metric foliation on a complete Riemannian manifold M, the holonomy maps h_c are still defined. The curve c is now a horizontal curve in M, and h_c is obtained by horizontally lifting projections of (restrictions of) c in local quotients. In general, however, h_c will only be a local diffeomorphism between leaves, since the latter need not share the same topology, cf. Examples and

Remarks 1.1.1(i). It should be noted that our definition of holonomy is fundamentally different from the one in [91]. There, the holonomy group of a leaf at a point p is understood to be a certain group of germs of diffeomorphisms of transverse manifolds that leave p invariant. Our definition, in contrast, is a generalization of the concept of holonomy in bundles. In fact, when the Riemannian submersion is a fiber bundle, and the horizontal distribution is a connection, we will later see that one recovers the usual holonomy group of the connection. If the submersion π is only a fibration, then the holonomy group will not, in general, be a Lie group: it will be shown in section 2.3 that if the holonomy group is a Lie group G, then π has the structure of a fiber bundle with group G. There are, however, examples of submersions whose structure group cannot be reduced to a Lie group [7], see also Remark 9.57 in [22]. It should furthermore be noted that even when it is a Lie group and M is compact, the holonomy group itself need not be compact [123].

(iv) Given a Riemannian submersion $\pi : M \to B$, horizontal lifts can in many cases easily be described directly. One such is the gradient ∇f of some function $f : B \to \mathbb{R}$ on the base. We claim that the basic lift of ∇f is $\nabla(f \circ \pi)$: Indeed, for any $e \in TM$,

$$\langle \nabla(f \circ \pi), e \rangle = e(f \circ \pi) = (\pi_* e)(f) = \langle \nabla f, \pi_* e \rangle. \tag{1.3.4}$$

Taking e to be vertical in (1.3.4) implies that $\nabla(f \circ \pi)$ is horizontal, and taking e to be horizontal shows that $\nabla(f \circ \pi)$ is π-related to ∇f.

In particular, if c is an integral curve of $\nabla(f \circ \pi)$, then

$$\dot{(\pi \circ \bar{c})} = \pi_* \circ \dot{c} = \pi_* \left(\nabla(f \circ \pi) \circ c\right) = \nabla f \circ (\pi \circ c),$$

so that $\pi \circ c$ is an integral curve of ∇f; equivalently, integral curves of $\nabla(f \circ \pi)$ are horizontal lifts of integral curves of ∇f.

(v) Let $p \in B$, and $\epsilon > 0$ smaller than the injectivity radius at p. The gradient of the distance function $f : B_\epsilon(p) \setminus \{p\} \longrightarrow \mathbb{R}$ from p, $f(q) := d(p, q)$, has as integral curves normal geodesics through p. Thus, (iv) yields a different proof of the existence of horizontal lifts of geodesics.

1.4 The fundamental tensors of a submersion

We are now ready to begin exploring the metric properties of a Riemannian submersion π. In this section, we will see that there are two tensor fields that measure the complexity of π; specifically, they determine by how much π differs from a projection $B \times F \to B$ of a metric product onto one of the factors. Given a Riemannian submersion $\pi : M \to B$, denote by ∇^M and ∇^B the Levi-Civita connections of M and B, and by R^M and R^B their curvature tensors.

Lemma 1.4.1. *If X, $Y \in \mathfrak{X}(M)$ are basic, then so is $(\nabla^M_X Y)^{\mathbf{h}}$. In fact, if $\tilde{X}$, $\tilde{Y}$ denote the vector fields on B that are π-related to X and Y, then $(\nabla^M_X Y)^{\mathbf{h}}$ is*

π-related to $\nabla^B_{\tilde X}\tilde Y$; i.e.,

$$\pi_*(\nabla^M_X Y)^{\mathbf{h}} = (\nabla^B_{\tilde X}\tilde Y)\circ\pi. \tag{1.4.1}$$

Proof. Given a basic $Z \in \mathfrak{X}(M)$ that is π-related to $\tilde Z \in \mathfrak{X}(B)$, the formula for the Levi-Civita connection implies

$$\begin{aligned} 2\langle \pi_*\nabla^M_X Y, \tilde Z\circ\pi\rangle &= 2\langle \pi_*\nabla^M_X Y, \pi_* Z\rangle = 2\langle \nabla^M_X Y, Z\rangle \\ &= X\langle Y,Z\rangle + Y\langle Z,X\rangle - Z\langle X,Y\rangle \\ &\quad + \langle Z,[X,Y]\rangle + \langle Y,[Z,X]\rangle - \langle X,[Y,Z]\rangle. \end{aligned}$$

Each term of the form $X\langle Y,Z\rangle$ above may be rewritten as

$$X\langle Y,Z\rangle = X\langle \pi_*Y,\pi_*Z\rangle = X(\langle\tilde Y,\tilde Z\rangle\circ\pi) = \pi_*X\langle\tilde Y,\tilde Z\rangle = (\tilde X\langle\tilde Y,\tilde Z\rangle)\circ\pi.$$

Similarly, $\langle Z,[X,Y]\rangle$ can be expressed as

$$\langle Z,[X,Y]\rangle = \langle\pi_*Z,\pi_*[X,Y]\rangle = \langle\tilde Z,[\tilde X,\tilde Y]\rangle\circ\pi.$$

Thus,

$$\begin{aligned} 2\langle \pi_*\nabla^M_X Y, \tilde Z\circ\pi\rangle &= \{\tilde X\langle\tilde Y,\tilde Z\rangle) + \tilde Y\langle\tilde Z,\tilde X\rangle - \tilde Z\langle\tilde X,\tilde Y\rangle \\ &\quad + \langle\tilde Z,[\tilde X,\tilde Y]\rangle + \langle\tilde Y,[\tilde Z,\tilde X]\rangle - \langle\tilde X,[\tilde Y,\tilde Z]\rangle\}\circ\pi \\ &= 2\langle\nabla^B_{\tilde X}\tilde Y,\tilde Z\rangle\circ\pi, \end{aligned}$$

which establishes (1.4.1). □

Notice that (1.4.1) also holds for metric foliations, since the argument is a local one.

Definition 1.4.1. The *A-tensor* of a metric foliation on M is the tensor field $A : \mathcal{H}\times\mathcal{H}\to\mathcal{V}$ on M given by

$$A_X Y = \nabla^{\mathbf{v}}_X Y = \frac{1}{2}[X,Y]^{\mathbf{v}}, \qquad X,Y\in\mathfrak{X}^{\mathbf{h}}. \tag{1.4.2}$$

The fact that A is indeed tensorial follows from (1.2.3).

Definition 1.4.2. The S-tensor of a metric foliation is the tensor field $S : \mathcal{H}\times\mathcal{V}\to\mathcal{V}$ on M given by

$$S_X U = -\nabla^{\mathbf{v}}_U X, \qquad X\in\mathfrak{X}^{\mathbf{h}},\quad U\in\mathfrak{X}^{\mathbf{v}}. \tag{1.4.3}$$

S_X is of course just the second fundamental tensor of a leaf in direction X. In particular, $S\equiv 0$ iff the leaves are totally geodesic (in which case we say the foliation is *totally geodesic*), whereas $A\equiv 0$ iff the horizontal distribution is integrable (in which case the foliation is said to be *flat*).

Remark 1.4.1. We caution the reader that the terminology we use here differs from that in [22] and [97] among others. There, the A-tensor is defined by the equation

$$A_E F = \nabla^{\mathbf{v}}_{E^{\mathbf{h}}} F^{\mathbf{h}} + \nabla^{\mathbf{h}}_{E^{\mathbf{h}}} F^{\mathbf{v}}.$$

Thus, the two notations coincide only for horizontal fields. Furthermore, these references use a tensor field T instead of S, defined by

$$T_E F = \nabla^{\mathbf{h}}_{E^{\mathbf{v}}} F^{\mathbf{v}} + \nabla^{\mathbf{v}}_{E^{\mathbf{v}}} F^{\mathbf{h}}.$$

For vertical fields U, V, $T_U V$ is the classical second fundamental tensor of a leaf, and is related to the S-tensor via

$$T_U V = \sum_i \langle S_{X_i} U, V\rangle X_i,$$

where X_i is a local orthonormal basis of horizontal fields.

The curvature formulae that will be derived in the next section involve covariant derivatives of these tensors, and for these to make sense, A and S must be defined for all vectors, not just horizontal or vertical ones. For this purpose, we set

$$A_E F = \nabla^{\mathbf{v}}_{E^{\mathbf{h}}} F^{\mathbf{h}}, \; S_E F = -\nabla^{\mathbf{v}}_{E^{\mathbf{v}}} F^{\mathbf{h}}, \qquad E, F \in \mathfrak{X},$$

and their covariant derivatives are then taken in the usual sense. For example,

$$(\nabla_E A)_{F_1} F_2 = \nabla_E (A_{F_1} F_2) - A_{\nabla_E F_1} F_2 - A_E \nabla_{F_1} F_2, \qquad E, F_i \in \mathfrak{X}.$$

The tensor fields A and S essentially determine the geometry of the metric foliation: The simplest example of a metric foliation is the one given by $\{b\} \times F$, $b \in B$, on a Riemannian product $M = B \times F$. A metric foliation is said to *split* if any point has a neighborhood isometric to a metric product, with the leaves tangent to one of the factors. Clearly, a foliation that splits is both totally geodesic and flat. Our next goal is to establish the converse.

Let us begin by taking a closer look at the holonomy transformations from Definition 1.3.1; recall that such a diffeomorphism $h_c : \pi^{-1}(b_0) \to \pi^{-1}(b_1)$ is generated by lifting a geodesic $c : [0,1] \to B$ joining points b_0 and b_1 in B to M: given $p \in \pi^{-1}(b_0)$, $h_c(p) = c_p(1)$, where c_p is the horizontal lift of c with $c_p(0) = p$. Denote by $A_x^* : \mathcal{V} \to \mathcal{H}$ the adjoint of $A_x : \mathcal{H} \to \mathcal{V}$:

$$\langle A_x^* u, y\rangle = \langle A_x y, u\rangle, \qquad x, y \in \mathcal{H}, \quad u \in \mathcal{V}.$$

Notice that if X is basic, then for vertical U,

$$A_X^* U = -\nabla^{\mathbf{h}}_U X, \tag{1.4.4}$$

since $[X, U]$ is vertical by (1.2.1), so that for horizontal Y,

$$\langle A_X^* U, Y\rangle = \langle A_X Y, U\rangle = \langle \nabla_X Y, U\rangle = X\langle Y, U\rangle - \langle Y, \nabla^{\mathbf{h}}_X U\rangle = \langle -\nabla^{\mathbf{h}}_U X, Y\rangle.$$

Lemma 1.4.2. *Let $\pi : M \to B$ denote a Riemannian submersion, and $h : F^0 \to F^1$ the holonomy diffeomorphism induced by the geodesic $c : [0,1] \to B$, where $c(0) = \pi(F^0)$, $c(1) = \pi(F^1)$. Given $p \in F^0$, let c_p denote the horizontal lift of c starting at p. Then for $u \in F^0_p$,*

$$h_* u = J(1), \tag{1.4.5}$$

where J is the Jacobi field along c_p with $J(0) = u$, $J'(0) = -A^*_{\dot c_p(0)} u - S_{\dot c_p(0)} u$.

Proof. Set $x := \dot c(0)$, and denote by X the basic field along F^0 with $\pi_* X = x$. Let $\gamma : I \to F^0$ be a curve defined on some neighborhood I of 0 with $\dot\gamma(0) = u$, and consider the variation $V : [0,1] \times I \to M$ of c_p given by $V(t,s) = \exp t(X \circ \gamma)(s)$. Then $h \circ \gamma(s) = V(1,s)$, so that $h_* u = V_* D_2(1,0)$. Since the variation is by geodesics, the vector field $t \mapsto J(t) = V_* D_2(t,0)$ is Jacobi along c, and $h_* u = J(1)$. Furthermore,

$$J'(0) = \nabla_{D_1} V_* D_2(0,0) = \nabla_{D_2} V_* D_1(0,0) = \nabla_u X,$$

and

$$\nabla_u X = \nabla^{\mathbf{h}}_u X + \nabla^{\mathbf{v}}_u X = -A^*_{\dot c_p(0)} u - S_{\dot c_p(0)} u$$

by (1.4.3) and (1.4.4). □

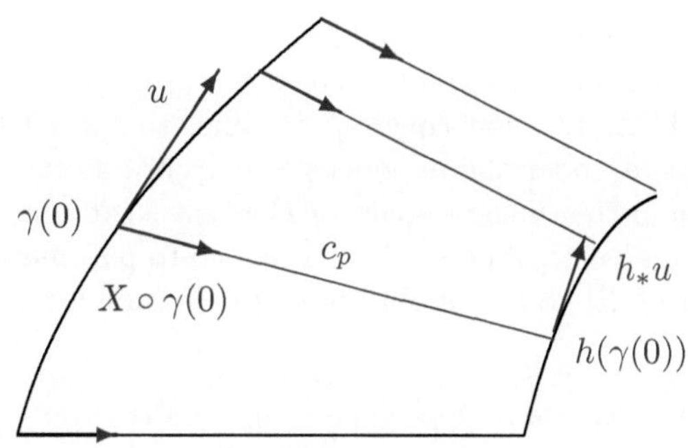

Definition 1.4.3. A Jacobi field J along a horizontal geodesic $c : [0,a] \to M$ that is vertical at 0 and satisfies $J'(0) = -A^*_{\dot c(0)} J(0) - S_{\dot c(0)} J(0)$ is called a *holonomy field*.

Such a field is always vertical, and is identically zero if it vanishes at one point, since the holonomy transformations are diffeomorphisms. Notice that for $t_0 \in (0,a)$, the restriction $J|_{[t_0,a]}$ is again a holonomy field along $c|_{[t_0,a]}$. Thus,

$$J'(t) = -(A^*_{\dot c(t)} + S_{\dot c(t)}) J(t), \qquad t \in [0,a]. \tag{1.4.6}$$

Lemma 1.4.3. *If $\pi : M \to B$ is a Riemannian submersion with totally geodesic fibers, and M is complete, then the holonomy diffeomorphisms between fibers are isometries.*

Proof. Consider a holonomy field J along a horizontal geodesic $c : [0,1] \to M$ with $J(0) = u$, as above. Since J is vertical, and the fibers are totally geodesic, (1.4.6) implies that

$$\langle J, J\rangle' = 2\langle J, J'\rangle = 2\langle J, J'^{\mathbf{v}}\rangle = -2\langle J, S_{\dot c} J\rangle = 0,$$

and J has constant norm. If h is the associated holonomy diffeomorphism, then by (1.4.5),

$$|h_* u| = |J(1)| = |J(0)| = |u|,$$

so that h is an isometry. □

Theorem 1.4.1. *Let $\pi : M \to B$ be a Riemannian submersion, with M complete.*

1. *If the fibers are totally geodesic, then $\pi : M \to B$ is a fiber bundle.*
2. *If, in addition, the A-tensor is identically zero, then π splits. Specifically, each point $b \in B$ has a neighborhood U such that $\pi^{-1}(U)$ is isometric to a metric product $U \times F$; furthermore, if $\Phi : U \times F \to \pi^{-1}(U)$ denotes the isometry, then, $\pi \circ \Phi : U \times F \to U$ is projection onto the first factor.*

Proof. We begin with the first statement, by constructing, as in [77], a principal bundle that has π as associated bundle. Fix $b_0 \in B$, let $F := \pi^{-1}(b_0)$, and G the Lie group of isometries of F. For any $b \in B$, G_b will denote the collection of all isometries $F \to \pi^{-1}(b)$. Define $P = \cup_{b\in B} G_b$. We claim that $\pi_P : P \to B$ admits a canonical G-bundle structure, where $\pi_P(h) = b$ if $h \in G_b$. There is a natural free right action of G on each set G_b: given $h \in G_b$, $g \in G$, set $hg := h \circ g : F \to \pi^{-1}(b)$. Let $\{U_\alpha\}$ be a locally finite open cover of B, such that each $U_\alpha = B_{\epsilon_\alpha}(b_\alpha)$ is the diffeomorphic image under the exponential map of an open ball of radius ϵ_α centered at the origin in the tangent space of B at b_α. Next, choose some geodesic c_α from b_0 to b_α. Given $b \in U_\alpha$, there is a unique (up to parametrization) geodesic c_b contained in U_α from b_α to b. We obtain in this fashion a map

$$\begin{aligned} s_\alpha : U_\alpha &\longrightarrow \pi_P^{-1}(U_\alpha), \\ b &\longmapsto h_{c_b} \circ h_{c_\alpha} \end{aligned}$$

that assigns to each $b \in U_\alpha$ the holonomy diffeomorphism (which in this case is an isometry by Lemma 1.4.3) along the piece-wise smooth curve c_α followed by c_b. Since the map

$$\begin{aligned} U_\alpha \times G &\longrightarrow \pi_P^{-1}(U_\alpha), \\ (b, g) &\longmapsto s_\alpha(b) g \end{aligned} \tag{1.4.7}$$

is a bijection, there exists a topology on P for which these maps become homeomorphisms.

It remains to check that the collection

$$\begin{aligned} (\pi_P, \phi_\alpha) : \pi_P^{-1}(U_\alpha) &\longrightarrow U_\alpha \times G, \\ h &\longmapsto (\pi_P(h), (s_\alpha \circ \pi_P(h))^{-1} \circ h) \end{aligned}$$

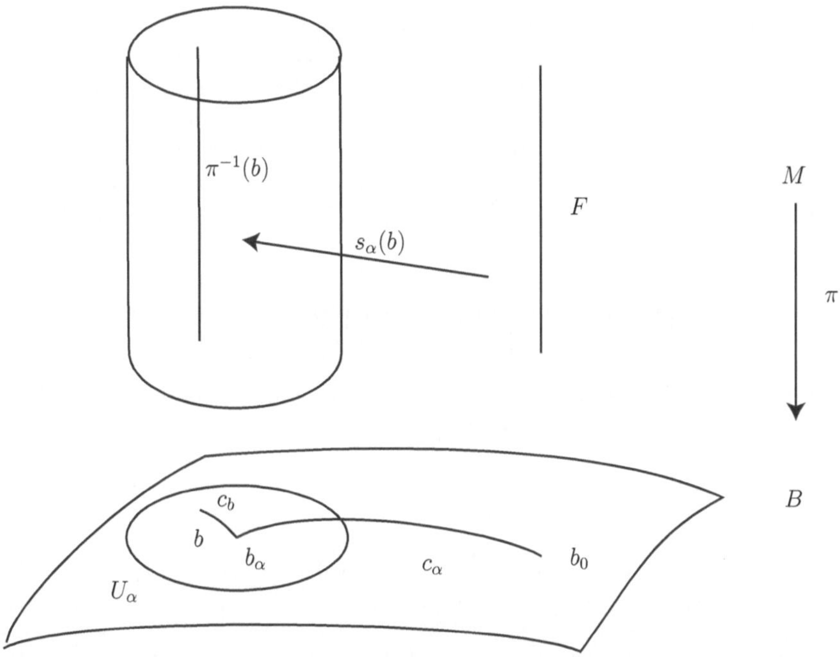

of inverses to (1.4.7) is a principal bundle atlas. But each ϕ_α is G-equivariant by construction, and if U_β intersects U_α, then

$$(\pi_P, \phi_\beta) \circ (\pi_P, \phi_\alpha)^{-1}(b, g) = (b, s_\beta(b)^{-1} \circ s_\alpha(b)g)$$

for $b \in U_\alpha \cap U_\beta$ and $g \in G$, so that the transition map $f_{\alpha,\beta} : U_\alpha \cap U_\beta \to G$ is given by $f_{\alpha,\beta}(b) = s_\beta(b)^{-1} \circ s_\alpha(b)$. Thus, $\pi_P : P \to B$ is a principal G-bundle over B. The diffeomorphism

$$\begin{aligned} P \times_G F &\longrightarrow M, \\ [h, q] &\longmapsto h(q) \end{aligned}$$

then exhibits $M \to B$ as the associated fiber bundle with fiber F.

To prove the second statement, let $b \in B$, $F := \pi^{-1}(b)$, and U a simply connected neighborhood of b that is the diffeomorphic image under $\exp_b$ of some ball in B_b. We consider again the local trivialization

$$\begin{aligned} (\pi, \phi) : \pi^{-1}(U) &\longrightarrow U \times F, \\ q &\longmapsto (\pi(q), c_q(1)) \end{aligned}$$

used in the proof of Theorem 1.3.1, where $c_q : [0, 1] \to F$ is the shortest geodesic from q to F. Notice that for any $\tilde{b} \in U$, the restriction $\phi|_{\pi^{-1}(\tilde{b})}$ is a holonomy diffeomorphism, and hence an isometry. We claim that for $p \in \pi^{-1}(U)$, and horizontal $x \in M_p$, $\phi_* x = 0$. Indeed, $\mathcal{H}$ is integrable, so that the restriction of π to the connected component $V \subset \pi^{-1}(U)$ of the leaf of $\mathcal{H}$ that contains p is a covering map, and therefore a diffeomorphism, U being simply connected. V then intersects F in exactly one point, namely $\phi(p)$. If γ is a horizontal curve with $\dot\gamma(0) = x$, then γ is contained in V, so that $\phi \circ \gamma = \phi(p)$, thereby establishing the claim. It follows that for $e \in M_p$, $|\pi_* e| = |e^{\mathbf{h}}|$, and $|\phi_* e| = |e^{\mathbf{v}}|$. Thus, $|(\pi, \phi)_* e| = |e|$, and (π, ϕ) is an isometry. □

We end this section by introducing a concept that will be needed in future sections:

Definition 1.4.4. The *mean curvature form* of a Riemannian foliation on M is the one-form κ on M given by $\kappa(E) = \operatorname{tr} S_{E^{\mathbf{h}}}$.

If T_i is a local orthonormal basis of the vertical space, then for horizontal X,

$$\kappa(X) = \sum_i -\langle T_i, \nabla_{T_i} X\rangle = \left\langle \sum_i \nabla_{T_i} T_i, X \right\rangle = \sum_i \langle T_i, [X, T_i]\rangle. \tag{1.4.8}$$

In particular, the *mean curvature vector field* H, given locally by $H = \sum_i \nabla^{\mathbf{h}}_{T_i} T_i$, is the vector field dual to κ.

Proposition 1.4.1. *For basic X, Y, $d\kappa(X,Y) = -2 \operatorname{div}(A_X Y)$.*

Proof. Since $d\kappa(X,Y) = X(\kappa Y) - Y(\kappa X) - \kappa[X,Y]$, the claim will follow once we establish that

$$\kappa([X,Y]) = \kappa([X,Y]^{\mathbf{h}}) = X(\kappa Y) - Y(\kappa X) + 2 \operatorname{div} A_X Y. \tag{1.4.9}$$

To see this, let $X_{ij} := \langle [X, T_i], T_j\rangle$, and $Y_{ij} := \langle [Y, T_i], T_j\rangle$. Since X is basic, each bracket $[X, T_i]$ is vertical, so that $[X, T_i] = \sum_j X_{ij} T_j$, and a similar formula holds for $[Y, T_i]$. Now,

$$\begin{aligned}
\kappa([X,Y]^{\mathbf{h}}) &= \sum_i \langle [[X,Y]^{\mathbf{h}}, T_i], T_i\rangle \\
&= \sum_i \{\langle [[X,Y], T_i], T_i\rangle - \langle [[X,Y]^{\mathbf{v}}, T_i], T_i\rangle\} \\
&= \sum_i \{\langle [[X,Y], T_i], T_i\rangle + \langle T_i, \nabla_{T_i} [X,Y]^{\mathbf{v}}\rangle\} \\
&= 2 \operatorname{div} A_X Y + \sum_i \langle [[X,Y], T_i], T_i\rangle,
\end{aligned}$$

since the divergence of a vertical field coincides with its divergence in a leaf.

Furthermore, by the Jacobi identity, the last term may be rewritten as follows:

$$\begin{aligned}\sum_i\langle[[X,Y],T_i],T_i\rangle &= \sum_i\{\langle[[X,T_i],Y],T_i\rangle - \langle[[Y,T_i],X],T_i\rangle\}\\ &= \sum_{i,j}\{\langle[X_{ij}T_j,Y],T_i\rangle - \langle[Y_{ij}T_j,X],T_i\rangle\}\\ &= \sum_{i,j}\{X_{ij}\langle[T_j,Y],T_i\rangle - Y(X_{ij})\langle T_j,T_i\rangle - Y_{ij}\langle[T_j,X],T_i\rangle\\ &\quad + X(Y_{ij})\langle T_j,T_i\rangle\}\\ &= \sum_{i,j}\{X_{ij}Y_{ji} - Y_{ij}X_{ji}\} + \sum_i\{X(Y_{ii}) - Y(X_{ii})\}\\ &= \sum_i\{X(Y_{ii}) - Y(X_{ii})\}\\ &= \sum_i\{X\langle[Y,T_i],T_i\rangle - Y\langle[X,T_i],T_i\rangle\}\\ &= X\kappa(Y) - Y\kappa(X). \qquad \square\end{aligned}$$

Examples and Remarks 1.4.1. (i) We will assume, unless specified otherwise, that the fiber F of a Riemannian submersion $\pi : M \to B$ is connected. By the long exact homotopy sequence of the fibration π, B is simply connected provided M is. If both fundamental tensors A and S vanish, and M is simply connected, de Rham's holonomy theorem implies that the local splitting guaranteed by Theorem 1.4.1 is actually global; i.e., M is isometric to the Riemannian product $B \times F$, cf. [22].

(ii) Although the results in this section are stated for submersions, they carry over as usual to foliations, at least locally. Thus, if a metric foliation has totally geodesic leaves, then the holonomy transformations are isometric, and if in addition the foliation is flat, then the ambient space splits locally as a metric product.

(iii) Suppose $\mathcal{F}$ is a metric foliation with totally geodesic leaves on M. Consider a leaf L, two horizontal vectors x and y at some point of L, and extend them to basic sections X, Y of the normal bundle of L in M. We claim that the vector field A_XY is a Killing field on L. In fact, if X is basic, then for vertical U,

$$\mathcal{L}_XU = (\mathcal{L}_XU)^{\mathbf{v}} = \nabla^{\mathbf{v}}_XU \tag{1.4.10}$$

since $[X,U]$ is vertical and $\nabla^{\mathbf{v}}_UX = -S_XU = 0$. Let U be a unit vertical field. We must show that $\langle\nabla_UA_XY,U\rangle = 0$, or equivalently, that

$$\langle\mathcal{L}_{[X,Y]^{\mathbf{v}}}U,U\rangle = 2(\langle\nabla_{A_XY}U,U\rangle - \langle\nabla_UA_XY,U\rangle) = -2\langle\nabla_UA_XY,U\rangle = 0. \tag{1.4.11}$$

By (1.4.10), $\langle\mathcal{L}_{[X,Y]^{\mathbf{h}}}U,U\rangle = 0$, so that

$$\begin{aligned}\langle\mathcal{L}_{[X,Y]^{\mathbf{v}}}U,U\rangle &= \langle\mathcal{L}_{[X,Y]}U,U\rangle = \langle\mathcal{L}_X\mathcal{L}_YU,U\rangle - \langle\mathcal{L}_Y\mathcal{L}_XU,U\rangle\\ &= \langle\nabla^{\mathbf{v}}_X\nabla^{\mathbf{v}}_YU - \nabla^{\mathbf{v}}_Y\nabla^{\mathbf{v}}_XU,U\rangle\\ &= \langle R(X,Y)U,U\rangle - \langle\nabla^{\mathbf{v}}_X\nabla^{\mathbf{h}}_YU - \nabla^{\mathbf{v}}_Y\nabla^{\mathbf{h}}_XU,U\rangle,\end{aligned}$$

and

$$\langle \mathcal{L}_{[X,Y]^{\mathbf{v}}} U, U \rangle = -\langle \nabla^{\mathbf{v}}_X \nabla^{\mathbf{h}}_Y U - \nabla^{\mathbf{v}}_Y \nabla^{\mathbf{h}}_X U, U \rangle. \tag{1.4.12}$$

Now, $\nabla^{\mathbf{v}}_X \nabla^{\mathbf{h}}_Y U = A_X \nabla^{\mathbf{h}}_Y U = -A_X A^*_Y U$. If T denotes the endomorphism $A_X A^*_Y$ of the vertical space, then $T^* = A_Y A^*_X$, and by (1.4.12),

$$\langle \mathcal{L}_{[X,Y]^{\mathbf{v}}} U, U \rangle = \langle (T - T^*)U, U \rangle = \langle TU, U \rangle - \langle U, TU \rangle = 0,$$

which establishes (1.4.11).

We will find in the next section a milder condition which ensures that $A_X Y$ is Killing along a leaf, see (1.5.12). It is already clear, though, that $A_X Y$ can be Killing without the leaves being totally geodesic: in fact, when the leaves are one-dimensional, the Killing condition is equivalent to $A_X Y$ having constant norm along leaves, and this is for example always satisfied in a space of constant curvature.

(iv) Let $\mathcal{F}$ be a metric foliation on a manifold M of nonpositive sectional curvature. If J is a Jacobi field along a geodesic c of M, then

$$\begin{aligned} |J|^{2\prime\prime} &= \langle J, J \rangle'' = 2\langle J', J \rangle' = 2(\langle J', J' \rangle + \langle J'', J \rangle) = 2(|J'|^2 - \langle R(J, \dot{c})\dot{c}, J \rangle) \\ &\geq 0. \end{aligned}$$

If $\mathcal{F}$ is totally geodesic, then the holonomy fields have constant norm, so that they must be parallel by the preceding inequality. (1.4.6) then implies that $A^*_{\dot{c}} J \equiv 0$ for a holonomy field along a horizontal geodesic c. Since $\dot{c}$ and J are arbitrary, $A \equiv 0$, and M splits locally as a metric product. In particular, a negatively curved manifold admits no totally geodesic metric foliations. A similar result holds if one removes the sectional curvature condition on M, and assumes instead that the leaves are compact and have negative Ricci curvature: A theorem of Bochner asserts that such a leaf cannot admit nontrivial Killing fields, and the claim then follows from (iii).

(v) Let $\mathcal{F}$ be a metric foliation on M. If $c : I \to M$ is an arbitrary curve, and E a vector field along c, then the covariant derivative E' of E may be expressed in terms of the fundamental tensors A and S. We will denote by σ the second fundamental tensor of the fibers,

$$\sigma(u, v) := \nabla^{\mathbf{h}}_U V, \qquad u, v \in M_p, \quad p \in M,$$

where U and V are local vertical extensions of u and v. Given $t_0 \in I$, consider a Riemannian submersion $\pi : U \to B$ defining $\mathcal{F}$ in a neighborhood U of $c(t_0)$, and extend $\dot{c}$ and E to vector fields F_c and F_E in U. Then $E^{\mathbf{v}\prime\mathbf{h}}$ is the restriction to c of

$$\nabla^{\mathbf{h}}_{F_c^{\mathbf{h}}} F^{\mathbf{v}}_E + \nabla^{\mathbf{h}}_{F_c^{\mathbf{v}}} F^{\mathbf{v}}_E = -A^*_{F_c^{\mathbf{h}}} F^{\mathbf{v}}_E + \sigma(F^{\mathbf{v}}_c, F^{\mathbf{v}}_E),$$

so that

$$E^{\mathbf{v}\prime\mathbf{h}} = -A^*_{\dot{c}^{\mathbf{h}}} E^{\mathbf{v}} + \sigma(\dot{c}^{\mathbf{v}}, E^{\mathbf{v}}). \tag{1.4.13}$$

Write $E^{\mathbf{h}} = \sum \phi^i (X_i \circ c)$, where X_i is a local basis of basic fields, and ϕ^i are functions defined on a neighborhood of t_0. Then

$$\begin{aligned} E^{\mathbf{h}/\mathbf{h}} &= \sum_i \phi^{i\prime} X_i \circ c + \sum \phi^i \nabla^{\mathbf{h}}_D (X_i \circ c) \\ &= \Big[\sum_i \phi^{i\prime} X_i \circ c + \sum_i \phi^i \nabla^{\mathbf{h}}_{\dot{c}^{\mathbf{h}}} X_i\Big] + \sum_i \phi^i \nabla^{\mathbf{h}}_{\dot{c}^{\mathbf{v}}} X_i, \end{aligned}$$

so that if $\tilde{x}$ denotes the horizontal lift of $x \in B_{\pi(p)}$ to M_p, then

$$E^{\mathbf{h}/\mathbf{h}} = \widetilde{(\pi_* E)'} - A^*_{E^{\mathbf{h}}} \dot{c}^{\mathbf{v}}. \tag{1.4.14}$$

Thus, the first term on the right side of the equation denotes the basic lift of the derivative of $\pi_* E$. Similarly,

$$\begin{aligned} E^{\mathbf{h}/\mathbf{v}} &= \sum_i \phi^i \nabla^{\mathbf{v}}_D (X_i \circ c) = \sum_i \phi^i \nabla^{\mathbf{v}}_{\dot{c}} X_i = \sum_i \phi^i \nabla^{\mathbf{v}}_{\dot{c}^{\mathbf{h}}} X_i + \sum_i \phi^i \nabla^{\mathbf{v}}_{\dot{c}^{\mathbf{v}}} X_i \\ &= \sum_i \phi^i A_{\dot{c}^{\mathbf{h}}} (X_i \circ c) - \sum_i \phi^i S_{X_i \circ c} \dot{c}^{\mathbf{v}}, \end{aligned}$$

and

$$E^{\mathbf{h}/\mathbf{v}} = A_{\dot{c}^{\mathbf{h}}} E^{\mathbf{h}} - S_{E^{\mathbf{h}}} \dot{c}^{\mathbf{v}}. \tag{1.4.15}$$

Equations (1.4.13)–(1.4.15) then yield

$$\begin{aligned} E'^{\mathbf{h}} &= \widetilde{(\pi_* E)'} - A^*_{E^{\mathbf{h}}} \dot{c}^{\mathbf{v}} - A^*_{\dot{c}^{\mathbf{h}}} E^{\mathbf{v}} + \sigma(\dot{c}^{\mathbf{v}}, E^{\mathbf{v}}), \\ E'^{\mathbf{v}} &= A_{\dot{c}^{\mathbf{h}}} E^{\mathbf{h}} - S_{E^{\mathbf{h}}} \dot{c}^{\mathbf{v}} + E^{\mathbf{v}/\mathbf{v}}. \end{aligned} \tag{1.4.16}$$

(1.4.16) provides another proof of the fact that $\mathcal{F}$ splits locally if A and S are identically zero. One can argue this using either horizontal or vertical fields: for example, if E is a parallel field along c with $E(t_0)$ vertical, then $E^{\mathbf{v}}$ is parallel, so $E = E^{\mathbf{v}}$ is always vertical. Thus, $\mathcal{V}$ is invariant under parallel translation, and the claim follows from de Rham's holonomy theorem.

(vi) If $\mathcal{F}$ is a metric foliation on M, then (1.4.16) applied to the tangent field $\dot{c}$ of a curve c in M yields

$$\dot{c}'^{\mathbf{h}} = \widetilde{(\pi_* \dot{c})'} - 2A^*_{\dot{c}^{\mathbf{h}}} \dot{c}^{\mathbf{v}} + \sigma(\dot{c}^{\mathbf{v}}, \dot{c}^{\mathbf{v}}), \qquad \dot{c}'^{\mathbf{v}} = -S_{\dot{c}^{\mathbf{h}}} \dot{c}^{\mathbf{v}} + \dot{c}^{\mathbf{v}/\mathbf{v}}. \tag{1.4.17}$$

Given a normal geodesic $c : I \to M$, define the *angle* $\measuredangle(\mathcal{F}, c) : I \to [0, \pi/2]$ *between* c *and* $\mathcal{F}$ by

$$\cos \measuredangle(\mathcal{F}, c) = |\dot{c}^{\mathbf{v}}|.$$

It is a differentiable function on I, since $\dot{c}^{\mathbf{v}} \equiv 0$ if it vanishes at one point. Suppose now that $\mathcal{F}$ is totally geodesic. By (1.4.17), $\dot{c}^{\mathbf{v}/\mathbf{v}} = 0$, so that $\langle \dot{c}^{\mathbf{v}}, \dot{c}^{\mathbf{v}} \rangle' = 2\langle \dot{c}^{\mathbf{v}/\mathbf{v}}, \dot{c}^{\mathbf{v}} \rangle = 0$. Thus if a metric foliation $\mathcal{F}$ is totally geodesic, then the geodesics of M make a constant angle with $\mathcal{F}$.

(vii) Denote by $\mathsf{H} : TM \longrightarrow \mathcal{H} \subset TM$ the orthogonal projection onto the horizontal subbundle. It is interesting to note that the A and S tensors are just the covariant derivatives in horizontal and vertical directions of this projection restricted to $\mathcal{H}$, cf. [80]. More precisely, for horizontal x and vertical u,

$$\nabla_x \mathsf{H}_{|\mathcal{H}} = A_x, \qquad S_x u = -(\nabla_u \mathsf{H})x.$$

In fact, for a horizontal field Y that evaluates to the vector y at the foot-point of x , $\nabla_x(\mathsf{H}Y) = (\nabla_x \mathsf{H})Y - \mathsf{H}\nabla_x Y$, so that $(\nabla_x \mathsf{H})Y = \nabla_x(\mathsf{H}Y) - \mathsf{H}\nabla_x Y = A_x y$, and a similar formula holds for $\nabla_u \mathsf{H}$. One can alternatively express this in terms of the covariant derivatives of the orthogonal projection $\mathsf{V} : TM \longrightarrow \mathcal{V}$ onto the vertical distribution, since $\mathsf{H} + \mathsf{V}$ is the identity, and the latter is parallel.

(viii) Recall that a map $f : M \longrightarrow N$ between Riemannian manifolds is said to be *harmonic* if it satisfies the Euler-Lagrange equation

$$\operatorname{tr} \nabla f_* = 0.$$

The covariant derivative operator in the above equation is the one induced by the Levi-Civita connections on M and N. Specifically, f_* is a section of the homomorphism bundle $\operatorname{Hom}(TM, f^*TN)$, so that

$$(\nabla_X f_*)Y = \nabla_X(f_* Y) - f_* \nabla_X Y, \qquad X, Y \in \mathfrak{X}M.$$

Suppose now that $\pi : M \longrightarrow B$ is a Riemannian submersion. Then for horizontal X, $(\nabla_X \pi_*)X = 0$ by (1.4.1), and for vertical T, $(\nabla_T \pi_*)T = -\pi_* \nabla_T T$. It follows that π is harmonic iff the mean curvature vector field is identically zero; i.e., iff the fibers are minimal submanifolds.

1.5 Curvature relations

Our next goal is to examine the relation between the curvature tensors R of M, R^B of B, and R^F of the fibers of a Riemannian submersion $\pi : M^{n+k} \to B^n$, $n \geq 2$. Denote by K and K^B the corresponding sectional curvatures, with $K_{x,y}$ denoting the curvature of the plane spanned by the vectors x and y. For the sake of brevity, we will omit these superscripts when dealing with the Levi-Civita connections. We begin with the horizontal curvatures. The following formula is commonly referred to as O'Neill's formula, even though it was also derived by Gray in [54].

Proposition 1.5.1. *For $p \in M$, and x, y, $z \in \mathcal{H}_p$,*

$$\pi_* R(x,y)z = R^B(\pi_* x, \pi_* y,)\pi_* z + \pi_*(2A_z^* A_x y - A_x^* A_y z - A_y^* A_z x).$$

Proof. Extend x, y, z locally to basic fields X, Y, Z, and denote by $\bar{X}$, $\bar{Y}$, $\bar{Z}$ the π-related vector fields on B. Then

$$\begin{aligned}\pi_* R(X,Y)Z &= \pi_*(\nabla_X \nabla_Y Z - \nabla_Y \nabla_X Z - \nabla_{[X,Y]} Z) \\ &= \pi_*(\nabla_X^{\mathbf{h}} \nabla_Y^{\mathbf{h}} Z + \nabla_X^{\mathbf{h}} \nabla_Y^{\mathbf{v}} Z - \nabla_Y^{\mathbf{h}} \nabla_X^{\mathbf{h}} Z - \nabla_Y^{\mathbf{h}} \nabla_X^{\mathbf{v}} Z \\ &\quad - \nabla_{[X,Y]^{\mathbf{h}}} Z - \nabla_{[X,Y]^{\mathbf{v}}} Z).\end{aligned} \tag{1.5.1}$$

Each term of the form $\pi_*(\nabla^{\mathbf{h}}_X \nabla^{\mathbf{h}}_Y Z)$ equals $(\nabla_{\bar X}\nabla_{\bar Y}\bar Z)\circ\pi$ by (1.4.1). Similarly,

$$\nabla^{\mathbf{h}}_X \nabla^{\mathbf{v}}_Y Z = \nabla^{\mathbf{h}}_X A_Y Z = -A^*_X A_Y Z,$$

and

$$\nabla^{\mathbf{h}}_{[X,Y]^{\mathbf{v}}} Z = -A^*_Z [X,Y]^{\mathbf{v}} = -2A^*_Z A_X Y$$

by (1.4.4). Substituting in (1.5.1) then establishes the claim. □

In order to investigate curvature relations involving one or more vertical vectors, recall from (1.4.6) that a holonomy Jacobi field J along a horizontal geodesic c satisfies

$$J' = -(A^*_{\dot c} + S_{\dot c})J.$$

As in Section 1.4, extend the tensor fields S and A to all of TM by setting

$$A_e f := A_{e^{\mathbf{h}}} f^{\mathbf{h}}, \quad S_e f := S_{e^{\mathbf{h}}} f^{\mathbf{v}}, \qquad e, f \in TM.$$

The covariant derivative $\nabla_D L$ of a tensor field L along c will be denoted by L'. Thus, for example, if L is of type $(1,1)$, then for a vector field E along c,

$$L'E = (LE)' - L(E').$$

Let T be a vertical vector field along c. Then

$$\begin{aligned}
\langle R(T,\dot c)\dot c, J\rangle &= \langle R(J,\dot c)\dot c, T\rangle = -\langle T, J''\rangle = \langle T, (A^*_{\dot c}J)'\rangle + \langle T, (S_{\dot c}J)'\rangle \\
&= \langle T, A_{\dot c}A^*_{\dot c}J\rangle + \langle T, S_{\dot c}J\rangle' - \langle T'^{\mathbf{v}}, S_{\dot c}J\rangle \\
&= \langle A_{\dot c}A^*_{\dot c}T, J\rangle + \langle S_{\dot c}T, J\rangle' - \langle S_{\dot c}(T'), J\rangle \\
&= \langle A_{\dot c}A^*_{\dot c}T, J\rangle + \langle (S_{\dot c}T)', J\rangle + \langle S_{\dot c}T, J'\rangle - \langle S_{\dot c}(T'), J\rangle \\
&= \langle A_{\dot c}A^*_{\dot c}T, J\rangle + \langle (S_{\dot c}T)', J\rangle - \langle S_{\dot c}(T'), J\rangle - \langle S^2_{\dot c}T, J\rangle \\
&= \langle (S'_{\dot c} - S^2_{\dot c} + A_{\dot c}A^*_{\dot c})T, J\rangle.
\end{aligned}$$

Given any t_0, the holonomy fields can be chosen so that they form an orthonormal basis of the vertical space at $c(t_0)$. Thus,

$$R^{\mathbf{v}}(T,\dot c)\dot c = (S'_{\dot c} - S^2_{\dot c} + A_{\dot c}A^*_{\dot c})T, \tag{1.5.2}$$

or equivalently,

$$R^{\mathbf{v}}(u,x)x = ((\nabla^{\mathbf{v}}_x S)_x - S^2_x + A_x A^*_x)u, \qquad x \in \mathcal{H}, \quad u \in \mathcal{V}. \tag{1.5.3}$$

Similarly, if X is a horizontal field along c, then

$$\langle R(X,\dot c)\dot c, J\rangle = -\langle X, J''\rangle = \langle X, (A^*_{\dot c}J)'\rangle + \langle X, (S_{\dot c}J)'\rangle.$$

The last term in the above equation equals $-\langle X, A_{\dot c}^* S_{\dot c} J\rangle = -\langle S_{\dot c} A_{\dot c} X, J\rangle$. The term before that may be rewritten as follows:

$$\begin{aligned}\langle X, (A_{\dot c}^* J)'\rangle &= \langle X, A_{\dot c}^* J\rangle' - \langle X', A_{\dot c}^* J\rangle = \langle A_{\dot c} X, J\rangle' - \langle A_{\dot c} X', J\rangle \\ &= \langle (A_{\dot c} X)', J\rangle + \langle A_{\dot c} X, -S_{\dot c} J\rangle - \langle A_{\dot c}(X'), J\rangle \\ &= \langle (A_{\dot c}'^{\mathbf{v}} - S_{\dot c} A_{\dot c}) X, J\rangle,\end{aligned}$$

so that

$$R^{\mathbf{v}}(X, \dot c)\dot c = (A_{\dot c}'^{\mathbf{v}} - 2 S_{\dot c} A_{\dot c}) X, \tag{1.5.4}$$

or equivalently,

$$R^{\mathbf{v}}(y, x)x = (\nabla_x^{\mathbf{v}} A)_x y - 2 S_x A_x y, \qquad x, y \in \mathcal{H}. \tag{1.5.5}$$

By polarization together with the Bianchi identity, (1.5.5) in turn implies

$$\begin{aligned}3R^{\mathbf{v}}(x, y)z &= R^{\mathbf{v}}(x, y+z)(y+z) - R^{\mathbf{v}}(y, x+z)(x+z) - R^{\mathbf{v}}(x, y)y \\ &\quad - R^{\mathbf{v}}(x, z)z + R^{\mathbf{v}}(y, x)x + R^{\mathbf{v}}(y, z)z \\ &= (\nabla_y^{\mathbf{v}} A)_z x - (\nabla_x^{\mathbf{v}} A)_z y + (\nabla_z^{\mathbf{v}} A)_y x - (\nabla_z^{\mathbf{v}} A)_x y \\ &\quad - 2 S_y A_z x + 2 S_x A_z y + 4 S_z A_x y.\end{aligned}$$

One easily checks that $\nabla_z^{\mathbf{v}} A$ is skew-symmetric; i.e.,

$$(\nabla_z^{\mathbf{v}} A)_x y = -(\nabla_z^{\mathbf{v}} A)_y x,$$

so that

$$3R^{\mathbf{v}}(x, y)z = (\nabla_y^{\mathbf{v}} A)_z x + (\nabla_x^{\mathbf{v}} A)_y z + 2(\nabla_z^{\mathbf{v}} A)_y x - 2 S_y A_z x + 2 S_x A_z y + 4 S_z A_x y. \tag{1.5.6}$$

In order to derive an alternative expression, we use the following:

Lemma 1.5.1. $\circlearrowleft (\nabla_x^{\mathbf{v}} A)_y z + \circlearrowleft S_x A_y z = 0$ *for* $x, y, z \in \mathcal{H}$, *where* $\circlearrowleft$ *denotes cyclic summation.*

Proof. Extend x, y, $z \in \mathcal{H}_p$ to basic fields X, Y, Z with vanishing horizontal Lie bracket at p. Then by the Jacobi identity,

$$\begin{aligned}0 &= \frac{1}{2} \circlearrowleft [X, [Y, Z]]^{\mathbf{v}} = \circlearrowleft [X, A_Y Z]^{\mathbf{v}} = \circlearrowleft \nabla_X^{\mathbf{v}}(A_Y Z) - \circlearrowleft \nabla_{A_Y Z}^{\mathbf{v}} X \\ &= \circlearrowleft \nabla_X^{\mathbf{v}}(A_Y Z) + \circlearrowleft S_X A_Y Z\end{aligned}$$

at p. It thus suffices to show that $\circlearrowleft (\nabla_X^{\mathbf{v}} A)_Y Z = \circlearrowleft \nabla_X^{\mathbf{v}}(A_Y Z)$. Now,

$$\begin{aligned}(\nabla_X^{\mathbf{v}} A)_Y Z - \nabla_X^{\mathbf{v}}(A_Y Z) &= -(A_{\nabla_X Y} Z + A_Y \nabla_X Z) = A_Z \nabla_X Y - A_Y \nabla_X Z \\ &= A_Z \nabla_X Y - A_Y \nabla_Z X\end{aligned}$$

since $[X, Z]$ is vertical. Cyclic summation of the last expression yields zero. $\square$

Applying Lemma 1.5.1 to the first two terms on the right of (1.5.6), we have

$$\begin{aligned}3R^{\mathbf{v}}(x,y)z &= \circlearrowleft (\nabla^{\mathbf{v}}_x A)_y z - (\nabla^{\mathbf{v}}_z A)_x y + 2(\nabla^{\mathbf{v}}_z A)_y x - 2S_y A_z x + 2S_x A_z y \\ &\quad + 4S_z A_x y \\ &= -3(\nabla^{\mathbf{v}}_z A)_x y - \circlearrowleft S_x A_y z - 2S_y A_z x + 2S_x A_z y + 4S_z A_x y \\ &= -3(\nabla^{\mathbf{v}}_z A)_x y + 3S_z A_x y - 3S_y A_z x - 3S_x A_y z,\end{aligned}$$

and

$$R^{\mathbf{v}}(x,y)z = -(\nabla^{\mathbf{v}}_z A)_x y + S_z A_x y - S_y A_z x - S_x A_y z, \qquad x,y,z \in \mathcal{H}. \tag{1.5.7}$$

We next compute $R^{\mathbf{v}}(x,u)y$ for horizontal x, y, and vertical u. As usual, extend these vectors to basic fields X, Y, and vertical field U. Then

$$\begin{aligned}R^{\mathbf{v}}(X,U)Y &= \nabla^{\mathbf{v}}_X \nabla^{\mathbf{h}}_U Y + \nabla^{\mathbf{v}}_X \nabla^{\mathbf{v}}_U Y - \nabla^{\mathbf{v}}_U \nabla^{\mathbf{h}}_X Y - \nabla^{\mathbf{v}}_U \nabla^{\mathbf{v}}_X Y - \nabla^{\mathbf{v}}_{[X,U]} Y \\ &= -\nabla^{\mathbf{v}}_X A^*_Y U - \nabla^{\mathbf{v}}_X S_Y U + S_{\nabla^{\mathbf{h}}_X Y} U - \nabla^{\mathbf{v}}_U A_X Y + S_Y [X,U]^{\mathbf{v}} \\ &= -A_X A^*_Y U - \nabla^{\mathbf{v}}_X S_Y U + S_Y \nabla^{\mathbf{v}}_X U + S_{\nabla^{\mathbf{h}}_X Y} U - S_Y \nabla^{\mathbf{v}}_U X \\ &\quad - \nabla^{\mathbf{v}}_U A_X Y \\ &= -A_X A^*_Y U - (\nabla^{\mathbf{v}}_X S)_Y U + S_Y S_X U - \nabla^{\mathbf{v}}_U (A_X Y).\end{aligned}$$

Everything but perhaps the last term is tensorial, and it can be rewritten by using the fact that

$$\begin{aligned}-(\nabla^{\mathbf{v}}_U A)_X Y &= -\nabla^{\mathbf{v}}_U (A_X Y) + A_{\nabla^{\mathbf{h}}_U X} Y + A_X \nabla^{\mathbf{h}}_U Y \\ &= -\nabla^{\mathbf{v}}_U (A_X Y) - A_{A^*_X U} Y - A_X A^*_Y U \\ &= -\nabla^{\mathbf{v}}_U (A_X Y) + A_Y A^*_X U - A_X A^*_Y U,\end{aligned} \tag{1.5.8}$$

which incidentally also shows that $(\nabla^{\mathbf{v}}_u A)_x y = -(\nabla^{\mathbf{v}}_u A)_y x$. Substituting this expression in the above equation for $R^{\mathbf{v}}(X,U)Y$ yields

$$R^{\mathbf{v}}(x,u)y = -(\nabla^{\mathbf{v}}_u A)_x y - A_y A^*_x u - (\nabla^{\mathbf{v}}_x S)_y u + S_y S_x u, \qquad x,y \in \mathcal{H}, \quad u \in \mathcal{V}. \tag{1.5.9}$$

Applying (1.5.9) to the right side of the identity

$$R^{\mathbf{v}}(x,y)u = R^{\mathbf{v}}(x,u)y - R^{\mathbf{v}}(y,u)x,$$

and recalling the skew-symmetry of $\nabla^{\mathbf{v}}_u A$, we obtain

$$R^{\mathbf{v}}(x,y)u = -2(\nabla^{\mathbf{v}}_u A)_x y - (\nabla^{\mathbf{v}}_x S)_y u + (\nabla^{\mathbf{v}}_y S)_x u + [S_y, S_x]u + (A_x A^*_y - A_y A^*_x)u, \tag{1.5.10}$$

where $[S_x, S_y] = S_x S_y - S_y S_x$. (1.5.10) may be expressed in a slightly more compact form using exterior covariant derivatives: recall that given a connection ∇ on a vector bundle ξ over M, the *exterior covariant derivative* of ∇ is the map d^{∇}

that assigns to each k-form $\omega \in A_k(M,\xi)$ on M with values in ξ the $(k+1)$-form $d^\nabla\omega \in A_{k+1}(M,\xi)$ given by

$$(d^\nabla\omega)(U_0,\dots,U_k) = \sum_{i=0}^{k}(-1)^i\nabla_{U_i}(\omega(U_0,\dots,\hat{U}_i,\dots,U_k)) + \sum_{i<j}(-1)^{i+j}\omega([U_i,U_j],U_0,\dots,\hat{U}_i,\dots,\hat{U}_j,\dots,U_k).$$

Now, the S-tensor may be viewed as a 1-form on M with values in the endomorphism bundle of TM. From this perspective, $d^\nabla S \in A_2(M,\operatorname{End}TM)$ is given by

$$d^\nabla S_{X,Y} = \nabla_X(S_Y) - \nabla_Y(S_X) - S_{[X,Y]^{\mathbf{h}}} = (\nabla_X S)_Y - (\nabla_Y S)_X,$$

and (1.5.10) then reads:

$$R^{\mathbf{v}}(x,y)u = -2(\nabla^{\mathbf{v}}_u A)_x y - d^\nabla S_{x,y}u + [S_y,S_x]u + (A_x A^*_y - A_y A^*_x)u.$$

For future reference, we group all these curvature identities together in the next theorem, along with the Gauss and Codazzi equations for the fibers of π:

Theorem 1.5.1 (Gray [54], O'Neill [97]). *Let $\pi : M^{n+k} \to B^n$ be a Riemannian submersion, $n \geq 2$, with R, R^B, and R^F denoting the curvature tensors of M, B, and a fiber F, respectively. Let $p \in M$, x, y, $z \in \mathcal{H}_p$, and u, v, $w \in \mathcal{V}$. Denote by σ the second fundamental tensor of the fiber $\pi^{-1}(\pi(p))$ at p, $\sigma(U,V) = \nabla^{\mathbf{h}}_U V$. Then*

$$\begin{aligned}
\pi_* R(x,y)z &= R^B(\pi_* x, \pi_* y,)\pi_* z + \pi_*(2A^*_z A_x y - A^*_x A_y z - A^*_y A_z x);\\
R^{\mathbf{v}}(x,y)z &= -(\nabla^{\mathbf{v}}_z A)_x y + S_z A_x y - S_y A_z x - S_x A_y z;\\
R^{\mathbf{v}}(x,u)y &= -(\nabla^{\mathbf{v}}_u A)_x y - A_y A^*_x u - (\nabla^{\mathbf{v}}_x S)_y u + S_y S_x u;\\
R^{\mathbf{v}}(x,y)u &= -2(\nabla^{\mathbf{v}}_u A)_x y - d^\nabla S_{x,y}u + [S_y,S_x]u + (A_x A^*_y - A_y A^*_x)u;\\
R^F(u,v)w &= R^{\mathbf{v}}(u,v)w + S_{\sigma(v,w)}u - S_{\sigma(u,w)}v;\\
R^{\mathbf{v}}(u,w)x &= (\nabla^{\mathbf{v}}_w S)_x u - (\nabla^{\mathbf{v}}_u S)_x w.
\end{aligned}$$

Recalling that the tensor field $\nabla^{\mathbf{v}}_u A$ is skew-symmetric, we immediately obtain for the sectional curvatures:

Corollary 1.5.1. *With notation as in Theorem 1.5.1, if x, y, u, v are orthonormal, then*

$$\begin{aligned}
K_{\pi_* x, \pi_* y} &= K_{x,y} + 3|A_x y|^2;\\
K^F_{u,v} &= K_{u,v} + \sigma(u,u)\sigma(v,v) - \sigma^2(u,v);\\
K_{x,u} &= \langle(\nabla^{\mathbf{v}}_x S)_x u, u\rangle + |A^*_x u|^2 - |S_x u|^2.
\end{aligned}$$

The second equation in the corollary is of course just the Gauss equation.

We mention one further consequence of the curvature identities above, which relates the derivatives of the S and A tensors. By (1.5.9),

$$\langle R^{\mathbf{v}}(x,u)y,v\rangle = -\langle(\nabla^{\mathbf{v}}_u A)_x y,v\rangle - \langle A_y A_x^* u,v\rangle - \langle(\nabla^{\mathbf{v}}_x S)_y u,v\rangle + \langle S_y S_x u,v\rangle.$$

Writing out the corresponding expression for $\langle R^v(y,v)x,u\rangle$, equating it to the one above, and noticing that the second and fourth term appear in both, we deduce

$$\langle(\nabla^{\mathbf{v}}_x S)_y u,v\rangle - \langle(\nabla^{\mathbf{v}}_y S)_x v,u\rangle = -\langle(\nabla^{\mathbf{v}}_u A)_x y,v\rangle + \langle(\nabla^{\mathbf{v}}_v A)_y x,u\rangle.$$

We have already remarked that $(\nabla^{\mathbf{v}}_u A)_y x = -(\nabla^{\mathbf{v}}_u A)_x y$. The above identity can therefore be rewritten as

$$\langle(\nabla^{\mathbf{v}}_y S)_x v,u\rangle - \langle(\nabla^{\mathbf{v}}_x S)_y u,v\rangle = \langle(\nabla^{\mathbf{v}}_u A)_x y,v\rangle + \langle(\nabla^{\mathbf{v}}_v A)_x y,u\rangle. \tag{1.5.11}$$

Extending x, y to basic fields X and Y, an alternative expression for the right side of (1.5.11) is $\langle\nabla_u(A_X Y),v\rangle + \langle\nabla_v(A_X Y),u\rangle$. This follows from (1.5.8) together with the fact that the operator $A_x A_y^* - A_y A_x^*$ is skew-adjoint. Furthermore, the operator $(\nabla^{\mathbf{v}}_x S)_y$ is self-adjoint, since

$$\begin{aligned}\langle(\nabla^{\mathbf{v}}_X S)_Y U,V\rangle &= \langle\nabla_X(S_Y U),V\rangle - \langle S_{\nabla^{\mathbf{h}}_X Y} U,V\rangle - \langle S_Y(\nabla^{\mathbf{v}}_X U),V\rangle\\ &= X\langle S_Y U,V\rangle - \langle S_Y U,\nabla^{\mathbf{v}}_X V\rangle - \langle\nabla^{\mathbf{h}}_U V,\nabla^{\mathbf{h}}_X Y\rangle\\ &\quad - \langle S_Y V,\nabla^{\mathbf{v}}_X U\rangle\end{aligned}$$

is symmetric in U and V. We then obtain the following alternative version of (1.5.11):

$$-\langle d^{\nabla} S_{x,y} u,v\rangle = \langle\nabla_u(A_X Y),v\rangle + \langle\nabla_v(A_X Y),u\rangle. \tag{1.5.12}$$

Examples and Remarks 1.5.1. (i) It follows from Corollary 1.5.1 that if $\pi : M \to B$ is a Riemannian submersion from a manifold M of nonnegative (resp. positive) sectional curvature, then the target space B also has nonnegative (resp. positive) curvature. It turns out that virtually all nonnegatively curved manifolds arise in this way. One typical example is that of projective spaces: the canonical metrics on $\mathbb{C}P^n$ and $\mathbb{H}P^n$ are those for which the natural projections $\pi : S^{2n+1} \to \mathbb{C}P^n$ and $\pi : S^{4n+3} \to \mathbb{H}P^n$ become Riemannian submersions. Consider for instance $\mathbb{C}P^n$. Denote by $\mathcal{J}_p : \mathbb{R}^{2n+2} \to \mathbb{R}^{2n+2}_p$ the canonical isomorphism sending the ith standard basis vector $\mathbf{e}_i$ to $D_i(p)$. Then the restriction N of the position vector field P of $\mathbb{R}^{2n+2}$, $P(p) = \mathcal{J}_p p$, to the unit sphere S^{2n+1} is a unit normal field to the sphere. Identify $\mathbb{R}^{2n+2}$ with $\mathbb{C}^{n+1}$ via

$$(x_1,y_1,\dots,x_{n+1},y_{n+1}) \mapsto (x_1+iy_1,\dots,x_{n+1}+iy_{n+1}),$$

and consider the canonical complex structure I on $T\mathbb{R}^{n+2}$ given by

$$I(\mathcal{J}_p v) = \mathcal{J}_p(iv), \qquad p,v \in \mathbb{C}^{n+1}.$$

The fiber of π through $(z_1,\dots,z_n)$ is $\{(z_1z,\dots,z_nz) \mid |z|=1\}$, so that IN is a unit vector field on the sphere that spans the fibers. Moreover, I is parallel, so that

$$\nabla_x IN = I\nabla_x N = Ix, \qquad x \in TS^{2n+1}. \tag{1.5.13}$$

The covariant derivatives in (1.5.13) are the Levi-Civita connection of Euclidean space, but since Ix and IN are both tangent to the sphere, the first covariant derivative also represents the Levi-Civita connection of the sphere.

The Hopf action of S^1 on S^{2n+1} is by isometries, so there exists a unique metric on complex projective space for which the Hopf fibration becomes a Riemannian submersion. Since IN is a unit field spanning the vertical distribution,

$$|A_xy|^2 = \langle A_xy, IN\rangle^2 = \langle y, A_x^*IN\rangle^2 = \langle y, (\nabla_x IN)^{\mathbf{h}}\rangle^2 = \langle y, Ix\rangle^2$$

for horizontal x and y. Here, we used the fact that if x is horizontal, then so is Ix: In fact, given any $x \in TS^{2n+1}$, $\langle Ix, IN\rangle = -\langle x, I^2N\rangle = \langle x, N\rangle = 0$. By Proposition 1.5.1,

$$K_{\pi_*x,\pi_*y} = 1 + 3\langle y, Ix\rangle^2$$

for orthonormal $x, y \in \mathcal{H}$. Thus, the sectional curvature K of $\mathbb{C}P^n$ satisfies $1 \le K \le 4$. For any horizontal x, the plane spanned by x and Ix projects down to a plane of curvature 4 (such a plane is sometimes called a *holomorphic* plane), whereas the plane spanned by x and any vector orthogonal to both x and Ix projects to a plane of curvature 1.

(ii) (An exotic sphere with nonnegative sectional curvature, [59]). Consider the Lie group $Sp(2)$ consisting of all 2×2 symplectic matrices; i.e., matrices Q with quaternion entries such that $QQ^* = Q^*Q = I_2$, where $Q^* = \bar{Q}^t$ denotes the transposed conjugate of Q. $Sp(2)$ admits a standard metric of nonnegative sectional curvature, namely the negative of its Killing form. The action of $Sp(1) \cong S^3$ on $Sp(2)$ given by

$$(q, Q) \longmapsto \begin{bmatrix} q & 0 \\ 0 & q \end{bmatrix} Q \begin{bmatrix} \bar{q} & 0 \\ 0 & 1 \end{bmatrix}$$

is a free action by isometries, so that there exists a unique Riemannian metric on the seven-dimensional quotient M^7 for which the projection $Sp(2) \to M^7$ becomes a Riemannian submersion. By Proposition 1.5.1, this metric has nonnegative curvature. We will see in 2.6.1 that M^7 is an exotic 7-sphere.

(iii) Since the curvature computations in this section are local in nature, Theorem 1.5.1 holds for metric foliations. Consider a metric foliation $\mathcal{F}$ on a complete space M^n of constant curvature κ. If $\kappa > 0$, then $\mathcal{F}$ cannot be flat anywhere: suppose, to the contrary, that $A_p \equiv 0$ for some $p \in M$. Let x be a unit horizontal vector at p, c the geodesic $t \mapsto \exp(tx)$ in direction x. If λ is an eigenvalue of the self-adjoint endomorphism S_x with corresponding eigenvector $u \in \mathcal{V}_p$, consider the holonomy field J along c with $J(0) = u$. Since

$$J'(0) = -A_x^*u - S_xu = -\lambda u = -\lambda J(0),$$

we have

$$J(t) = \left(\cos\sqrt{\kappa}t - \frac{\lambda}{\sqrt{\kappa}}\sin\sqrt{\kappa}t\right)E(t),$$

where E is the parallel field along c with $E(0) = u$. But then J vanishes somewhere, and is therefore identically zero, contradicting $u \neq 0$. The same argument shows that $A_x \neq 0$ for any nonzero horizontal x.

Suppose next that $\kappa = 0$. We claim that if $A_p \equiv 0$, then $A \equiv 0$ everywhere, and the foliation splits. In fact, arguing as above, we see that holonomy fields along horizontal geodesics emanating from p must be parallel: if $u \in \mathcal{V}_p$ is a λ-eigenvector of S_x, $x \in \mathcal{H}_p$, then the holonomy field J along c with $J(0) = u$ is given by

$$J(t) = (1 - \lambda t)E(t),$$

where E is the parallel field along c with $E(0) = u$. Since J cannot vanish, $\lambda = 0$, and J is parallel. It follows that if N denotes the totally geodesic submanifold $\exp\mathcal{H}_p$, then the vertical distribution is orthogonal to N everywhere; i.e., $A \equiv 0$ and $S \equiv 0$ along N. But since M has constant curvature, $A \equiv 0$ along the leaf through p, and both A and S must then vanish everywhere.

(iv) Consider the isometric $\mathbb{R}$-action on Euclidean 3-space given by glide rotations:

$$(t, (z, t_0)) \longmapsto (e^{it}z, t + t_0), \qquad t \in \mathbb{R}, \quad (z, t_0) \in \mathbb{C} \times \mathbb{R} = \mathbb{R}^3.$$

This action is free and there exists a unique metric of nonnegative sectional curvature on the two-dimensional quotient M^2 (diffeomorphic to $\mathbb{R}^2$) for which the projection $\pi : \mathbb{R}^3 \to M^2$ is a Riemannian submersion. This metric is rotationally symmetric, and we compute its curvature in terms of the distance $r = (x^2 + y^2)^{1/2}$ to the z-axis; i.e., $K(r)$ will denote the sectional curvature of M at $\pi(F)$, where F is an orbit in $\mathbb{R}^3$ at distance r from the z-axis. The vertical space at $(x + iy, t_0)$ is spanned by $\dot{c}(0)$, where $c(t) = (e^{it}(x + iy), t + t_0)$, so that a unit vertical vector field is given by

$$T = \frac{1}{(1 + r^2)^{1/2}}(yD_1 - xD_2 + D_3).$$

Thus,

$$X = \frac{1}{r}(xD_1 + yD_2), \qquad Y = \frac{1}{r(1 + r^2)^{1/2}}(-yD_1 + xD_2 + r^2D^3)$$

is an orthonormal basis of horizontal vector fields away from the z-axis. A lengthy computation now yields

$$[X, Y] = \frac{1 + 2r^2}{r^2(1 + r^2)^{3/2}}(yD_1 - xD_2) + \frac{1}{(1 + r^2)^{3/2}}D_3,$$

and by Corollary 1.5.1,

$$K(r) = \frac{3}{4}\langle [X, Y], T\rangle^2 = \frac{3}{(1 + r^2)^2}.$$

(v) Taking $u = v$ in (1.5.12) and summing over an orthonormal basis of the vertical space implies that for basic X, Y,

$$2 \operatorname{div} A_X Y = -\operatorname{tr} d^\nabla S_{X,Y}.$$

This also follows directly from Proposition 1.4.1, which asserts that the mean curvature κ of the foliation satisfies $-d\kappa(X,Y) = 2 \operatorname{div} A_X Y$: since κ is just the trace of S, the above identity then follows from $\operatorname{tr} \circ d^\nabla = d \circ \operatorname{tr}$, which itself reflects the fact that the trace operator is a parallel section of the bundle $(\operatorname{End} TM)^*$, see for example [136].

(vi) The reader may have noticed that some of the curvature formulae in this section seem to differ substantially from those in [22] or [97], even allowing for the difference in notation. For example, O'Neill's formula for $\langle R(x,y)u, v\rangle$ does not involve the derivative of the S-tensor, as ours does. However, substituting (1.5.12) in our expression for $R^{\mathrm{v}}(x,y)u$ yields

$$\begin{aligned}\langle R(x,y)u,v\rangle &= \langle(\nabla_v A)_x y, u\rangle - \langle(\nabla_u A)_x y, v\rangle + \langle[S_y, S_x]u, v\rangle \\ &\quad + \langle(A_x A_y^* - A_y A_x^*)u, v\rangle,\end{aligned}$$

which agrees with O'Neill's formula.

1.6 Projectable Jacobi fields

We have seen that holonomy fields along a horizontal geodesic $c : [0,a] \to M$ arise by lifting a single geodesic, namely $\pi \circ c$, horizontally along some curve in the fiber through $c(0)$. In this section, we consider a larger class of Jacobi fields along c; roughly speaking, these fields arise by horizontally lifting a variation of $\pi \circ c$ rather than only $\pi \circ c$. Let $\mathcal{F}$ denote a metric foliation on M, $c : [0,a] \to M$ a horizontal geodesic. A Jacobi field J along c is said to be *projectable* if it satisfies

$$J'^{\mathrm{v}} = -S_{\dot c} J^{\mathrm{v}} - A_{\dot c} J^{\mathrm{h}}. \tag{1.6.1}$$

The collection of projectable Jacobi fields along c is clearly a vector space that contains the collection of holonomy fields as a subspace.

Let I denote an interval containing 0, and for $s \in I$, denote by $\imath_s : [0,a] \to [0,a] \times I$ the map sending t to (t,s). The *variational field* of a variation $V : [0,a] \times I \to M$ of c is defined to be the vector field $V_* D_2 \circ \imath_0$ along c. Recall that if V is a variation by geodesics (meaning that $V_s := V \circ \imath_s$ is a geodesic for every $s \in I$), then its variational field is Jacobi.

Proposition 1.6.1. *If $V : [0,a] \times I \to M$ is a variation of c through horizontal geodesics, then its variational field is projectable.*

Proof. Fix $t_0 \in [0, a]$, and set $\gamma(s) := V(t_0, s)$, $E(s) := V_*D_1(t_0, s)$. Then

$$\begin{aligned} J'^{\mathbf{v}}(t_0) &= \nabla^{\mathbf{v}}_{D_1} V_* D_2(t_0, 0) = \nabla^{\mathbf{v}}_{D_2} V_* D_1(t_0, 0) = E'^{\mathbf{v}}(0) \\ &= A_{\dot\gamma^{\mathbf{h}}} E^{\mathbf{h}}(0) - S_{E^{\mathbf{h}}} \dot\gamma^{\mathbf{v}}(0) + E^{\mathbf{v}\prime\mathbf{v}}(0), \end{aligned}$$

by (1.4.16). But E is by assumption horizontal, so that

$$J'^{\mathbf{v}}(t_0) = A_{J^{\mathbf{h}}} \dot c(t_0) - S_{\dot c} J^{\mathbf{v}}(t_0) = -(S_{\dot c} J^{\mathbf{v}} + A_{\dot c} J^{\mathbf{h}})(t_0). \qquad \square$$

Establishing the converse to Proposition 1.6.1 requires a little more work:

Lemma 1.6.1. *Let $\pi : M \to B$ denote a Riemannian submersion, $c : [0, a] \to M$ a horizontal geodesic, and J a Jacobi field along $\pi \circ c$. Given any $u \in \mathcal{V}_{c(0)}$, there exists a unique projectable Jacobi field $\tilde J$ along c such that*

$$(1) \quad \pi_* \tilde J = J, \qquad \textit{and} \qquad (2) \quad \tilde J^{\mathbf{v}}(0) = u.$$

Proof. Let I be a small interval around 0, $\gamma : I \to B$ a curve with $\gamma(0) = c(0)$, $\dot\gamma(0) = J(0)$. If V, W denote the parallel fields along γ with $V(0) = \dot c(0)$, $W(0) = J'(0)$, consider the variation $V : [0, a] \times I \to B$ of $\pi \circ c$ given by

$$V(t, s) = \exp_{\gamma(s)} t(V + sW)(s).$$

It is easily checked that $J = V_* D_2 \circ \imath_0$. Next, let $\bar\gamma : I \to M$ be a curve with $\pi \circ \bar\gamma = \gamma$, whose initial tangent vector has u as vertical component. For each $s \in I$, denote by X_s the basic field along the fiber through $\gamma(s)$ with $\pi_* X_s = (V + sW)(s)$, and consider the variation $\tilde V : [0, a] \times I \to M$ of c given by

$$\tilde V(t, s) = \exp_{\bar\gamma(s)} t(X_s \circ \bar\gamma)(s).$$

Since the latter is by horizontal geodesics, its variational field $\tilde J$ is a projectable Jacobi field by Proposition 1.6.1, and $\tilde J^{\mathbf{v}}(0) = u$ by construction. Moreover, $\pi \circ \tilde V = V$, so that $\pi_* \tilde J = \pi_* \tilde V_* D_2 \circ \imath_0 = V_* D_2 \circ \imath_0 = J$. By (1.6.1),

$$\tilde J^{\mathbf{v}\prime\mathbf{v}} = \tilde J'^{\mathbf{v}} - A_{\dot c} \tilde J^{\mathbf{h}} = -S_{\dot c} \tilde J^{\mathbf{v}} - 2A_{\dot c} \tilde J^{\mathbf{h}},$$

so that

$$\tilde J^{\mathbf{v}\prime} = -(S_{\dot c} + A^*_{\dot c}) \tilde J^{\mathbf{v}} - 2A_{\dot c} \tilde J^{\mathbf{h}}.$$

This, together with (2), determines $\tilde J^{\mathbf{v}}$ uniquely. But $\tilde J^{\mathbf{h}}$ is determined by (1), and uniqueness of $\tilde J$ follows. $\square$

Theorem 1.6.1. *If $\pi : M \to B$ is a Riemannian submersion, and J is a projectable Jacobi field along a horizontal geodesic $c : [0, a] \to M$, then $\pi_* J$ is Jacobi along $\pi \circ c$.*

Proof. Denote by $\mathcal{P}$ the space of projectable Jacobi fields along c and by $\mathcal{J}$ the space of all Jacobi fields along $\pi \circ c$. By Lemma 1.6.1, the map

$$\begin{aligned} \mathcal{V}_{c(0)} \times \mathcal{J} &\longrightarrow \mathcal{P}, \\ (u, J) &\longmapsto \bar{J} \end{aligned} \tag{1.6.2}$$

that sends the pair (u, J) to the unique projectable Jacobi field $\bar{J}$ with $\pi_* \bar{J} = J$ and $\bar{J}^{\mathbf{v}}(0) = u$ is well defined and linear. Its kernel is trivial by construction, so that $\dim \mathcal{P} \geq k + 2 \dim B = \dim M + \dim B$ (here, k is the fiber dimension). Next, consider the linear map

$$\begin{aligned} \mathcal{P} &\longrightarrow M_{c(0)} \times \mathcal{H}_{c(0)}, \\ J &\longmapsto (J(0), J'^{\mathbf{h}}(0)). \end{aligned}$$

If J is an element in the kernel, then $J'^{\mathbf{h}}(0) = 0$, and $J'^{\mathbf{v}}(0) = -S_{\dot{c}(0)} J^{\mathbf{v}}(0) - A_{\dot{c}(0)} J^{\mathbf{h}}(0) = 0$. Thus, $J(0) = J'(0) = 0$, and the kernel is trivial. This implies $\dim \mathcal{P} \leq \dim M + \dim B$, and by the reverse inequality above, $\dim \mathcal{P} = \dim M + \dim B$. The 1-1 map from (1.6.2) is therefore an isomorphism, and the theorem follows. □

Corollary 1.6.1. *Let $\mathcal{F}$ be a metric foliation on M. If J is a projectable Jacobi field along a horizontal geodesic $c : [0, a] \to M$, then there exists a variation of c by horizontal geodesics that has J as its variational field.*

Proof. Since $c[0, a]$ is compact and connected, we may assume that $\mathcal{F}$ is defined by a Riemannian submersion π. By Theorem 1.6.1, $\pi_* J$ is Jacobi along $\pi \circ c$, and the claim then follows from the proof of Lemma 1.6.1, where such a variation is explicitly constructed. □

Examples and Remarks 1.6.1. (i) A Jacobi field along $c : [0, a] \to M$ that is projectable at one point $t_0 \in [a, b]$ is projectable on all of $[a, b]$. In fact, the space $\mathcal{P}_{t_0}$ of Jacobi fields along c that satisfy (1.6.1) at the point t_0 contains the set $\mathcal{P}$ of projectable Jacobi fields as a subspace, and has the same dimension as $\mathcal{P}$: as in the proof of Theorem 1.6.1, the map

$$\begin{aligned} \mathcal{P}_{t_0} &\longrightarrow M_{c(t_0)} \times \mathcal{H}_{c(t_0)}, \\ J &\longmapsto (J(t_0), J'^{\mathbf{h}}(t_0)) \end{aligned}$$

is an isomorphism.

(ii) Recall that for a submanifold F of M, and a normal geodesic $c : I = [0, a] \to M$ with $\dot{c}(0) \perp F_{c(0)}$, a point t_0 in I is said to be a *focal point* of F along c if there exists a nontrivial Jacobi field J along c with $J(0) \in F_{c(0)}$, $J'^{\mathbf{v}}(0) = -S_{\dot{c}(0)} J(0)$, and $J(t_0) = 0$. Here $u^{\mathbf{v}}$ denotes the orthogonal projection of $u \in M_{c(0)}$ onto $F_{c(0)}$, and S is the second fundamental tensor of F. Suppose $\pi : M \to B$ is a Riemannian submersion, $c : [0, a] \to M$ a normal horizontal geodesic, and

$F = \pi^{-1}(\pi(c(0)))$. By Theorem 1.6.1, if $t_0 \in I$ is a focal point of F along c, then t_0 is a conjugate point of $\pi \circ c$: in fact, the Jacobi field J above is projectable at 0, hence everywhere by (i). $\pi_* J$ is then a Jacobi field that vanishes at 0 and t_0. It turns out that the order of focal points of c and conjugate points of $\pi \circ c$, as well as the indexes of these geodesics, are the same, see [98] for details and further results.

1.7 The Riccati equation for Jacobi fields

When dealing with Jacobi fields, it is often useful to decompose the second-order Jacobi differential equation into two first-order ones. Let V denote a vector space of Jacobi fields orthogonal to a geodesic $c : \mathbb{R} \to M^n$. We assume that V is self-adjoint in the sense that $\langle J_1, J_2' \rangle(t) = \langle J_1', J_2 \rangle(t)$ for any $J_i \in \mathsf{V}$ and some (and hence all) $t \in \mathbb{R}$. Given $t \in \mathbb{R}$, set

$$V(t) := \{J(t) \mid J \in \mathsf{V}\} \oplus \{J'(t) \mid J \in \mathsf{V}, J(t) = 0\} \subset M_{c(t)}.$$

$V(t)$ is clearly a subspace of $M_{c(t)}$.

Lemma 1.7.1. $\dim V(t) = \dim \mathsf{V}$. *Furthermore, the second summand is trivial for almost every* t.

Proof. Notice first that the sum is indeed a direct one, since the summands are mutually orthogonal: If $J_i(t)$ belongs to the ith summand, $i = 1, 2$, then $\langle J_1(t), J_2'(t) \rangle = \langle J_1'(t), J_2(t) \rangle = 0$. Next, set

$$V_1(t) := \{J \in \mathsf{V} \mid J(t) = 0\}, \qquad V_2(t) := \{J'(t) \mid J \in \mathsf{V}, J(t) = 0\}.$$

The map $V_1(t) \to V_2(t)$ which sends $J \in V_1(t)$ to $J'(t)$ is linear, surjective by definition, and has zero kernel. Thus $V_1(t) \cong V_2(t)$. For any fixed $t_0 \in \mathbb{R}$, let $J_1, \dots, J_k$ be a basis of $V_1(t_0)$, and extend it to a basis $J_1, \dots, J_l$ of V, where $l := \dim \mathsf{V}$. Then $J_{k+1}(t_0), \dots, J_l(t_0)$ is a basis of $\{J(t_0) \mid J \in \mathsf{V}\}$, and

$$\begin{aligned} \dim \mathsf{V} = l = k + l - k &= \dim V_1(t_0) + \dim\{J(t_0) \mid J \in \mathsf{V}\} \\ &= \dim V_2(t_0) + \dim\{J(t_0) \mid J \in \mathsf{V}\} = \dim V(t_0). \end{aligned}$$

To prove the second statement, let t_0 and $J_1, \dots, J_l$ be as above. We will show that there is some $\epsilon > 0$ such that $J_1(t), \dots, J_l(t)$ are linearly independent for each $t \in (t_0, t_0 + \epsilon)$. By assumption, $J_{k+1}(t_0), \dots, J_l(t_0)$ are linearly independent, and thus remain so for t close enough to t_0. On the other hand, if $\epsilon > 0$ is such that the restriction of c to $[t_0, t_0 + \epsilon]$ has no conjugate points, then $J_1(t), \dots, J_k(t)$ must also be linearly independent for each $t \in (t_0, t_0 + \epsilon)$: Indeed, if $\sum_{i \le k} \alpha_i J_i(t) = 0$ for some $t \in (t_0, t_0 + \epsilon)$, then $J := \sum \alpha_i J_i$ vanishes at t_0 and t, so that $J \equiv 0$. It now remains to show that if $J \in \operatorname{span}\{J_1, \dots, J_k\}$, then $J(t) \notin \operatorname{span}\{J_{k+1}(t), \dots, J_l(t)\}$. To see this, write $J = \sum f_i E_i$, where $E_1, \dots, E_{n-1}$ are orthonormal parallel fields perpendicular to c, and $f_i = \langle J, E_i\rangle$. Since $f_i(t_0) = 0$, $f_i(t) = (t - t_0)g_i(t)$ for some function g_i with $g_i(t_0) = f_i'(t_0)$. Thus,

$$\frac{1}{t-t_0}J(t) \to J'(t_0) \perp \operatorname{span}\{J_i(t_0) \mid i > k\} \text{ as } t \to t_0,$$

which establishes the claim. □

Suppose next that $\dim \mathsf{V} = n - 1$, where n is the dimension of M. Then for almost every t, $\dot c^{\perp}(t)$ is spanned by $\{J(t) \mid J \in \mathsf{V}\}$, and we may define a self-adjoint operator $S(t)$ on $\dot c^{\perp}(t)$ by setting $S(t)u := J'(t)$, where J is the element of V with $J(t) = u$. S is called the *Riccati operator* of V. Let $R(t)$ denote the self-adjoint operator $R(\cdot, \dot c(t))\dot c(t)$ on $\dot c^{\perp}(t)$. Then for $J \in \mathsf{V}$,

$$-R(t)J(t) = J''(t) = (SJ)'(t) = (S'J)(t) + (SJ')(t) = (S' + S^2)(t)J(t).$$

Identifying $\dot c^{\perp}(t)$ with $W := \dot c^{\perp}(0)$ via parallel translation along c, each $J \in \mathsf{V}$ is a curve in W, and S, R are curves in the space $\mathsf{S}(W)$ of self-adjoint transformations of W. Furthermore, the second-order Jacobi equation along c is now decomposed into two first-order equations

$$SJ = J', \qquad S' + S^2 + R = 0. \tag{1.7.1}$$

There exists a comparison theory for the above equation which can be used to derive the Rauch comparison theorems for Jacobi fields, cf. [44], [45]. Here, we shall only consider a special case. Endow $\mathsf{S}(W)$ with the inner product $\langle A, B\rangle = \operatorname{tr}(AB)$. Taking traces of the Riccati equation $S' + S^2 + R = 0$ yields

$$(\operatorname{tr} S)' + \operatorname{tr}(S^2) + \operatorname{Ric} = 0, \tag{1.7.2}$$

with $\operatorname{Ric} = \operatorname{tr} R$. Decompose S as $S = (\operatorname{tr} S/(n-1))I + S_0$, where S_0 is the traceless part of S, so that $\langle I, S_0\rangle = 0$. Then

$$\operatorname{tr}(S^2) = |S|^2 = \frac{(\operatorname{tr} S)^2}{n-1} + |S_0|^2,$$

and setting $s := \operatorname{tr} S/(n-1)$, (1.7.2) becomes

$$s' + s^2 + r = 0, \qquad r = \frac{\operatorname{Ric} + |S_0|^2}{n-1}. \tag{1.7.3}$$

Theorem 1.7.1. *Let V be an $(n-1)$-dimensional space of Jacobi fields orthogonal to $c : \mathbb{R} \to M^n$ with self-adjoint Riccati operator S. Suppose furthermore, that $\{J(t) \mid J \in \mathsf{V}\}$ spans $\dot c(t)^{\perp}$ for all $t \in \mathbb{R}$. If $\operatorname{Ric}(\dot c) \geq 0$, then $S \equiv 0$, and $\operatorname{Ric}(\dot c) \equiv 0$. In particular, V consists of parallel Jacobi fields.*

Proof. Since $\{J(t) \mid J \in \mathsf{V}\}$ spans $\dot{c}(t)^\perp$ for all $t \in \mathbb{R}$, the Riccati operator S is defined on all of $\mathbb{R}$. Suppose, to the contrary, that S is not identically zero, and consider the associated equation (1.7.3). Then $s(t_0) \neq 0$ for some $t_0 \in \mathbb{R}$ (for if $s \equiv 0$, then $r \equiv 0$, hence $S \equiv S_0 \equiv 0$). We may, without loss of generality, assume that $t_0 = 0$ and that $s(0) < 0$ (otherwise, consider $t \mapsto \tilde{s}(t) = -s(-t + t_0)$). Let f denote the solution of

$$f' + f^2 = 0, \qquad f(0) = s(0);$$

i.e.,

$$f(t) = \frac{1}{t + 1/s(0)}.$$

If $y := f - s$ on $(-\infty, -1/s(0))$, then $y' = f' - s' = -f^2 + s^2 + r = -(f-s)(f+s) + r$, so that y is the solution of the O.D.E.

$$y' = -(f+s)y + r, \qquad y(0) = 0.$$

Consider any nontrivial solution x of the homogeneous equation $x' = -(1/2)(f+s)x$. If u is the function satisfying $u' = 2r/x^2$, $u(0) = 0$, then $y = (1/2)ux^2$. But $u \geq 0$, so that $y \geq 0$, and $s \leq f$. This contradicts the fact that s is defined for all time, since $f(t) \to -\infty$ as $t \nearrow -1/s(0)$. □

The condition that the Jacobi fields in V span the normal space of c at every point is essential: Consider for example $\mathbb{R}^2$ with the standard metric, and a geodesic c in $\mathbb{R}^2$. If E is a parallel field orthogonal to c, then $t \mapsto J(t) := tE(t)$ defines a non-parallel Jacobi field, and represents an $(n-1)$-dimensional space of Jacobi fields satisfying all the conditions of the theorem except for the one above. The corresponding Riccati operator is $S(t) = (1/t)I$ which is not defined at 0.

When $\dim \mathsf{V} < n - 1$, one can still, following Wilking [141], derive a Riccati-type equation for Jacobi fields that are transversal to V: Suppose J is an $(n-1)$-dimensional space of Jacobi fields orthogonal to c with self-adjoint Riccati operator, and V is a subspace of J. As before, define

$$V(t) = \{J(t) \mid J \in \mathsf{V}\} \oplus \{J'(t) \mid J \in \mathsf{V}, J(t) = 0\}.$$

By Lemma 1.7.1, $V(t)$ and V have the same dimension, and the second summand vanishes almost everywhere. Let $H(t) := V(t)^\perp \cap \dot{c}(t)^\perp$, and write $u = u^{\mathbf{v}} + u^{\mathbf{h}} \in V(t) \oplus H(t)$. Given a generic $t_0 \in \mathbb{R}$ (i.e., one for which $\dot{c}(t_0)^\perp$ is spanned by $\{J(t_0) \mid J \in \mathsf{J}\}$), there is a well-defined operator $S(t_0) : H(t_0) \to H(t_0)$ given by

$$S(t_0)u := Y'^{\mathbf{h}}(t_0), \text{ where } Y = J^{\mathbf{h}},\ J \in \mathsf{J}, \text{ and } J(t_0) = u.$$

To see this, notice that if $J_i(t_0) = u$ for $i = 1, 2$, then for any $J \in \mathsf{J}$,

$$\langle (J_1 - J_2)', J\rangle(t_0) = \langle J_1 - J_2, J'\rangle(t_0) = 0,$$

so that $J_1'(t_0) = J_2'(t_0)$, and hence $J_1'^{\mathbf{h}}(t_0) = J_2'^{\mathbf{h}}(t_0)$. Furthermore, $J_i^{\mathbf{v}\prime\mathbf{h}}(t_0) = 0$, because $J_i^{\mathbf{v}}(t_0) = 0$, so that given a vector field X tangent to H, $\langle J_i^{\mathbf{v}\prime}, X\rangle(t_0) = -\langle J_i^{\mathbf{v}}, X'\rangle(t_0) = 0$. Thus, $J_1^{\mathbf{h}\prime\mathbf{h}}(t_0) = J_2^{\mathbf{h}\prime\mathbf{h}}(t_0)$ as claimed. Next, we check that S is self-adjoint: let $u_i = Y_i(t_0) \in H(t_0)$, where $Y_i = J_i^{\mathbf{h}}$, $J_i(t_0) = u_i$. As remarked above, $J_i^{\mathbf{v}\prime\mathbf{h}}(t_0) = 0$, so that

$$\begin{aligned}\langle S(t_0)u_1, u_2\rangle &= \langle J_1^{\mathbf{h}\prime}, J_2\rangle(t_0) = \langle (J_1 - J_1^{\mathbf{v}})', J_2\rangle(t_0) = \langle J_1', J_2\rangle(t_0)\\ &= \langle J_1, J_2'\rangle(t_0) = \langle u_1, S(t_0)u_2\rangle.\end{aligned}$$

Now, the assignment that sends a vector field X tangent to H to $X'^{\mathbf{h}}$ determines a covariant derivative operator $D^{\mathbf{h}}$ on the bundle $\{(t,u) \mid t \in \mathbb{R}, u \in H(t)\}$ over $\mathbb{R}$, and therefore also a system of parallel translations. Use the latter to identify $H(t)$ with $E := H(t_0)$, and obtain in the process a map $S : I \to \mathsf{S}(E)$ from some interval I containing t_0 into the space of self-adjoint operators on E. With this identification, $S' = D^{\mathbf{h}}S$, and for $Y = J^{\mathbf{h}}$,

$$Y'' = D^{\mathbf{h}2}Y = D^{\mathbf{h}}(SY) = (D^{\mathbf{h}}S)Y + S(D^{\mathbf{h}}Y) = (S' + S^2)Y. \tag{1.7.4}$$

But $D^{\mathbf{h}2}Y$ can also be computed as follows: Define $A(t) : V(t) \to H(t)$ by $A(t)u = J'^{\mathbf{h}}(t)$, where $J \in \mathsf{V}$, $J(t) = u$. As we did for S, it is straightforward to check that $A(t)$ is a well-defined linear map. As usual, assume that $Y(t_0) = J(t_0)$ for $Y = J^{\mathbf{h}}$ as above, and consider orthonormal $D^{\mathbf{h}}$-parallel fields $X_1, \dots X_d$ with $X_1(t_0) = J(t_0)$. Notice that for $Z \in \mathsf{V}$,

$$\langle J', Z\rangle(t_0) = \langle J, Z'\rangle(t_0) = \langle J, AZ\rangle(t_0),$$

so that

$$J'^{\mathbf{v}}(t_0) = A^*(t_0)J(t_0). \tag{1.7.5}$$

Similarly,

$$X_i'(t) = -A^*(t)X_i(t), \tag{1.7.6}$$

since

$$0 = \langle X_i, Z\rangle' = \langle X_i', Z\rangle + \langle X_i, Z'^{\mathbf{h}}\rangle = \langle X_i', Z\rangle + \langle X_i, AZ\rangle = \langle X_i' + A^*X_i, Z\rangle,$$

and $X_i'^{\mathbf{h}} \equiv 0$. Thus, setting $R := R(\cdot)\dot{c}, \dot{c}$, (1.7.5) and (1.7.6) imply

$$\begin{aligned}\langle D^{\mathbf{h}2}Y, X_k\rangle(t_0) &= \langle J^{\mathbf{h}\prime\mathbf{h}\prime}, X_k\rangle(t_0) = \langle J^{\mathbf{h}\prime\mathbf{h}}, X_k\rangle'(t_0)\\ &= \langle J^{\mathbf{h}}, X_k\rangle''(t_0) - \langle J^{\mathbf{h}}, X_k'\rangle'(t_0) = \langle J, X_k\rangle''(t_0)\\ &= \langle J'', X_k\rangle(t_0) + 2\langle J', X_k'\rangle(t_0) + \langle J, X_k''\rangle(t_0)\\ &= -\langle RJ, X_k\rangle(t_0) - 2\langle A^*J, A^*X_k\rangle(t_0) + \langle X_1, X_k''\rangle(t_0)\\ &= -\langle RY, X_k\rangle(t_0) - 2\langle AA^*J, X_k\rangle(t_0) - \langle X_1', X_k'\rangle(t_0)\\ &\quad + \langle X_1, X_k'\rangle'(t_0)\\ &= -\langle RY - 3AA^*Y, X_k\rangle(t_0),\end{aligned}$$

and

$$D^{h2}Y + R^{\mathbf{h}}Y + 3AA^*Y = 0.$$

Together with (1.7.4), we conclude

$$S' + S^2 + R^{\mathbf{h}} + 3AA^* = 0. \tag{1.7.7}$$

Theorem 1.7.2 (Wilking, [141]). *Let J denote an $(n-1)$-dimensional space of Jacobi fields orthogonal to a geodesic $c : \mathbb{R} \to M^n$, with self-adjoint Riccati operator. If M has nonnegative sectional curvature, then*

$$\mathsf{J} = \mathrm{span}_{\mathbb{R}}\{J \in \mathsf{J} \mid J(t) = 0 \textit{ for some } t\} \oplus \{J \in \mathsf{J} \mid J \textit{ is parallel }\}.$$

Proof. Set

$$\mathsf{V} := \mathrm{span}_{\mathbb{R}}\{J \in \mathsf{J} \mid J(t) = 0 \text{ for some } t\}.$$

The collection $\{Y \mid Y = J^{\mathbf{h}}, J \in \mathsf{J}\}$ is a vector space canonically isomorphic to J/V via $J + \mathsf{V} \mapsto J^{\mathbf{h}}$, and satisfies (1.7.7). The argument used in the proof of Theorem 1.7.1 can now be repeated verbatim to deduce that S in (1.7.7) identically vanishes, provided we know that this collection spans $H(t)$ for every t; equivalently, that for any $J \in \mathsf{J} \setminus \mathsf{V}$, and any $t \in \mathbb{R}$, the vector $J(t)$ is transversal to

$$V(t) = \{J(t) \mid J \in \mathsf{V}\} \oplus \{J'(t) \mid J \in \mathsf{V}, J(t) = 0\}.$$

Suppose, to the contrary, that for some $J \in \mathsf{J} \setminus \mathsf{V}$ and some t, $J(t) = J_1(t) + J_2'(t)$, $J_i \in \mathsf{V}$, $J_2(t) = 0$. Then $\langle J(t), J_2'(t)\rangle = \langle J'(t), J_2(t)\rangle = 0$, and similarly, $J_1(t) \perp J_2'(t)$. But then $J_2'(t) = J(t) - J_1(t)$ is perpendicular to itself, hence vanishes. Thus, $J(t) = J_1(t)$, and $J - J_1 \in \mathsf{V}$, which contradicts the fact that $J \notin \mathsf{V}$. By (1.7.7), $S \equiv 0$, and the fields Y are $D^{\mathbf{h}}$-parallel. But A (and $R^{\mathbf{h}}$) also vanish, so that $Y' = -A^*Y = 0$, and any Y is then parallel for the Levi-Civita connection. □

If $\pi : M^{n+k} \to B^n$ is a metric fibration, and $c : \mathbb{R} \to M$ is a horizontal geodesic in M, then there is a distinguished $(n + k - 1)$-dimensional family of projectable Jacobi fields orthogonal to c with self-adjoint Riccati operator that will be used in the next section: define

$$\mathcal{P}_v = \{J \mid J \text{ is projectable Jacobi along } c, \quad J \perp \dot{c}, \quad J^{\mathbf{h}}(0) = 0\}. \tag{1.7.8}$$

An argument similar to the one used in Theorem 1.6.1 shows that $\mathcal{P}_v$ has dimension $n + k - 1$. Denote by $\pi_*\mathcal{P}_v$ the space of projected Jacobi fields π_*J, $J \in \mathcal{P}_v$. Since every element in $\pi_*\mathcal{P}_v$ vanishes at 0, $\pi_*\mathcal{P}_v$ has self-adjoint Riccati operator. The following lemma then implies that $\mathcal{P}_v$ itself also has that property:

Lemma 1.7.2. *Let $\pi : M \to B$ be a metric fibration, $c : \mathbb{R} \to M$ a horizontal geodesic, and V a subspace of projectable Jacobi fields along c. If $\pi_*\mathsf{V}$ denotes the space $\{\pi_*J \mid J \in \mathsf{V}\}$ of projected Jacobi fields, then V has self-adjoint Riccati operator iff $\pi_*\mathsf{V}$ does.*

Proof. Given J_1 and J_2 in V,

$$\begin{aligned}\langle J_1, J_2'\rangle &= \langle J_1^{\mathbf{v}}, J_2'^{\mathbf{v}}\rangle + \langle J_1^{\mathbf{h}}, J_2'^{\mathbf{h}}\rangle \\ &= \langle J_1^{\mathbf{v}}, -S_{\dot c}J_2^{\mathbf{v}}\rangle - \langle J_1^{\mathbf{v}}, A_{\dot c}J_2^{\mathbf{h}}\rangle + \langle J_1^{\mathbf{h}}, J_2^{\mathbf{h}\prime\mathbf{h}}\rangle + \langle J_1^{\mathbf{h}}, J_2^{\mathbf{v}\prime\mathbf{h}}\rangle \\ &= \langle J_1^{\mathbf{v}}, -S_{\dot c}J_2^{\mathbf{v}}\rangle - \langle A_{\dot c}^* J_1^{\mathbf{v}}, J_2^{\mathbf{h}}\rangle + \langle J_1^{\mathbf{h}}, J_2^{\mathbf{h}\prime\mathbf{h}}\rangle - \langle J_1^{\mathbf{h}}, A_{\dot c}^* J_2^{\mathbf{v}}\rangle.\end{aligned}$$

Thus,

$$\begin{aligned}\langle J_1, J_2'\rangle - \langle J_1', J_2\rangle &= \langle J_1^{\mathbf{h}}, J_2^{\mathbf{h}\prime\mathbf{h}}\rangle - \langle J_1^{\mathbf{h}\prime\mathbf{h}}, J_2^{\mathbf{h}}\rangle \\ &= \langle \pi_* J_1, (\pi_* J_2)'\rangle - \langle (\pi_* J_1)', \pi_* J_2\rangle,\end{aligned}$$

and the claim follows. □

1.8 The dual foliation

Consider a metric foliation $\mathcal{F}$ on a Riemannian manifold M that may be singular (see Section 1.2) in the sense that the leaves need not have constant dimension. Wilking [141] introduced a so-called *dual foliation* of M that plays a key role in nonnegative sectional curvature: given $p \in M$, the dual leaf through p is defined by

$$L^{\#}(p) := \{q \in M \mid \text{there exists a piece-wise smooth horizontal curve from } p \text{ to } q\}.$$

To show that the collection $\mathcal{F}^{\#}$ of dual leaves is indeed a (singular) foliation, it must be checked that these leaves are smooth immersed submanifolds of M. For a fixed leaf L of $\mathcal{F}$ with normal bundle $\pi : \nu(L) \to L$, consider all vector fields X on M that can be written as $\exp_{\nu *} Y \circ \exp_\nu^{-1}$ for some smooth vector field Y with compact support on $\nu(L)$ such that $\pi_* Y = 0$. The latter condition means that the points on a flow line of X can be connected to a single point in L by means of horizontal geodesics. In particular, flow lines are contained in a dual leaf.

Next, consider the collection $\mathfrak{X}_0$ of all vector fields obtained in this way from each and every leaf of $\mathcal{F}$, and the group $\mathcal{D}$ of diffeomorphisms of M generated by the flows of these fields. Finally, denote by $\mathfrak{X}$ the Lie algebra generated by all vector fields induced by one-parameter subgroups of $\mathcal{D}$, and set

$$\Delta_p := \{X(p) \mid X \in \mathfrak{X}\}, \qquad p \in M.$$

In general, Δ will not be a distribution globally, because it need not have constant dimension, cf. Examples and Remarks 1.8.1(iii). Notice, however, that if $X \in \mathfrak{X}$ has flow Φ_t and $\phi \in \mathcal{D}$, then $\phi_* \circ X \circ \phi^{-1}$ has flow $\phi \circ \Phi_t \circ \phi^{-1}$, and hence also belongs to $\mathfrak{X}$. Thus, Δ has constant dimension along orbits of $\mathcal{D}$. It follows that the orbits are precisely the integral manifolds of this distribution. Each orbit is therefore a smooth submanifold, which by definition coincides with a dual leaf.

If k denotes the dimension of a given leaf of the original foliation, then the dimension of a dual leaf $L^\#$ that intersects it must of course satisfy $\dim M - k \le \dim L^\# \le \dim M$. Both bounds may be taken on: The simplest example of a dual foliation occurs for a Riemannian product $M = B \times F$. The metric projection $M \to B$ is a Riemannian submersion, and the dual foliation consists of the collection $B \times \{q\}$, as q ranges over F. On the other hand, it is not hard to see that for the Hopf fibration $S^3 \to S^2$, the dual foliation has exactly one leaf; i.e., any two points of S^3 can be joined by a piece-wise smooth horizontal curve.

Wilking [141] has shown that this is actually a property shared by all foliations in spaces of positive curvature:

Theorem 1.8.1. *The dual of a metric foliation in a space M^{n+k} of positive curvature consists of a single leaf.*

Proof. By definition, horizontal geodesics are contained in dual leaves. The theorem follows once we show that the normal space to a leaf along some geodesic contained in it is spanned by parallel Jacobi fields. Suppose then that some dual leaf $L^\#$ has dimension smaller than $n := \dim M$, and consider a horizontal geodesic c in $L^\#$. If the tangent space of $L^\#$ at $c(0)$ intersects the vertical space, choose an orthonormal basis $u_1, \dots, u_k$ of $\mathcal{V}_{c(0)}$ with $u_l, \dots, u_k \in L^\#_{c(0)}$, $l > 1$. By hypothesis, there exists for each $i \ge l$ a curve γ_i with $\dot\gamma_i(0) = u_i$, such that if s is small enough, then $\gamma_i(s)$ can be joined to $c(0)$ by a broken horizontal geodesic. Thus, there exist Jacobi fields $J_l, \dots, J_k$ with $J_i(0) = u_i$, and $J_i(t) \in L^\#_{c(t)}$. Similarly, choose Jacobi fields $J_{k+1}, \dots, J_{k+n-1}$ with $J_{k+i}(0) = 0$, such that $\{J'_{k+i}(0)\}$ spans $\mathcal{H}_{c(0)} \cap \dot c(0)^\perp$. By Corollary 1.6.1, $J_{k+1}(t), \dots, J_{k+n-1}(t)$ are also tangent to $L^\#$ for all t. Finally, denote by $J_1, \dots, J_{l-1}$ the holonomy fields with $J_i(0) = u_i$. Now, $\mathsf{J} := \operatorname{span}_{\mathbb{R}}\{J_1, \dots, J_{n+k-1}\}$ is an $(n+k-1)$-dimensional space of projectable Jacobi fields orthogonal to c that are vertical at 0. By dimension considerations, it coincides with the space $\mathcal{P}_v$ from (1.7.8). Thus, J has self-adjoint Riccati operator, and by Theorem 1.7.2,

$$\mathsf{J} = \operatorname{span}_{\mathbb{R}}\{J \in \mathsf{J} \mid J(t) = 0 \text{ for some } t\} \oplus \{J \in \mathsf{J} \mid J \text{ is parallel }\}.$$

Notice that the first summand above is contained in $\{J_i \in \mathsf{J} \mid i \ge l\}$: For if $J \in \mathsf{J}$, then J is a projectable Jacobi field, so that if $J(t_0) = 0$ for some t_0, it is the variational field of a variation by horizontal geodesics emanating from $c(t_0)$; i.e., $J(t) \in L^\#_{c(t)}$ for all t. Thus, J contains a subspace of parallel Jacobi fields of dimension $\ge l-1$. Since the two summands of J are point-wise orthogonal, the normal space of $L^\#$ along c is spanned by parallel Jacobi fields, as claimed. □

Notice that the proof does not require positive curvature everywhere, so that the theorem actually holds on any space of nonnegative curvature where the curvature is strictly positive at one point and dual leaves are complete. Wilking's result is related to so-called Carnot-Carathéodory structures studied by Gromov and others, which have deep applications in rigidity problems for noncompact rank one symmetric spaces, cf. [92], [100], [65]. Such a structure is defined by a subbundle $\mathcal{H}$ of the tangent bundle of a manifold M together with an inner product, so that a horizontal curve (i.e., one tangent to $\mathcal{H}$) has a well-defined length. One then sets the distance between two points to be the infimum of the lengths of all horizontal curves that connect them (if such curves exist), or ∞ otherwise. If any two points can be joined by a horizontal curve, then this distance is actually a metric on M, called a Carnot-Carathéodory or sub-Riemannian metric, and this metric induces the original topology on M. Theorem 1.8.1 says that in a space M of positive curvature, the horizontal distribution stemming from a metric foliation defines a Carnot-Carathéodory metric on M. In general, if one only assumes the existence of a horizontal distribution $\mathcal{H}$ on a manifold M that is not necessarily endowed with a Riemannian structure, then Chow's connectivity theorem asserts that any two points can be joined by a horizontal curve provided $\mathcal{H}$ is completely nonintegrable; i.e., the tangent space of M at any point can be generated by taking iterated brackets of horizontal vector fields.

Examples and Remarks 1.8.1. (i) The proof of Theorem 1.8.1 does not require that the dimension k of a leaf be constant. This implies that the theorem holds more generally for singular Riemannian foliations in positively curved manifolds.

(ii) By definition, the distribution spanned by the dual foliation contains the image of the A-tensor. It is therefore easy to find examples of metric foliations in spaces that are not positively curved where any two points can be joined by a broken horizontal geodesic: one such is the foliation of $\mathbb{R}^3$ given by the orbits of the isometric $\mathbb{R}$-action

$$\begin{aligned} \mathbb{R} \times (\mathbb{R}^2 \times \mathbb{R}) &\to \mathbb{R}^2 \times \mathbb{R}, \\ (t, (p, t_0)) &\mapsto (e^{it} p, t + t_0). \end{aligned}$$

If, on the other hand, we allow singular Riemannian foliations, then there are even examples of flat foliations with only one dual leaf: perhaps the simplest one consists of the foliation of $\mathbb{R}^2$ by circles of radius r around a point, with $0 \le r < \infty$.

(iii) When the dual foliation consists of more than one leaf, it is in general singular. One example that will be treated in more detail in Chapter 3 is the orbit space $M = S^3 \times_{S^1} \mathbb{R}^2$ of the free isometric action of S^1 on the Riemannian product $S^3 \times \mathbb{R}^2$, where each factor has the standard metric; the action of $z \in S^1$ on $(p, u) \in S^3 \times \mathbb{R}^2$ is given by $z(p, u) := (pz^{-1}, zu)$, with $p \in S^3$ viewed as an element of $\mathbb{C}^2$, and $u \in \mathbb{C}$. Since the action is by isometries, there is a metric of nonnegative curvature on M for which the projection $\rho : S^3 \times \mathbb{R}^2 \to M$ is a Riemannian submersion. Notice that $\rho(S^3 \times \{0\}) = S^2(1/2)$ is the image of the Hopf fibration. Consider the map

$$\begin{aligned} \pi : M = \rho(S^3 \times \mathbb{R}^2) &\to \rho(S^3 \times \{0\}), \\ \rho(p, u) &\mapsto \rho(p, 0). \end{aligned}$$

If $\mathsf{p}_1 : S^3 \times \mathbb{R}^2 \to S^3$ denotes projection onto the first factor, and $\sigma : S^3 \to S^2(1/2)$ is the Hopf fibration, then $\sigma \circ \mathsf{p}_1 = \pi \circ \rho$. Furthermore, $(x, 0) \in S^3_p \times \mathbb{R}^2_u$ is

ρ-horizontal whenever $x \in S_p^3$ is σ-horizontal, so that $|\pi_* \circ \rho_*(x,0)| = |\sigma_* x| = |x|$, and π is Riemannian. This also shows that any ρ-horizontal curve c with $c(0) = (p,u)$ has its image contained inside $\rho(S^3 \times \{u\})$, and in fact, the restriction of π to the positively curved manifold $\rho(S^3 \times \{u\})$ is again Riemannian. By Theorem 1.8.1, the foliation dual to π consists of three-dimensional leaves $\rho(S^3 \times \{u\})$, $|u| > 0$, and one two-dimensional leaf $S^2(1/2)$.

(iv) Wilking has shown that if the dual foliation has complete leaves, then it is also a (singular) metric foliation provided the ambient space is nonnegatively curved, see [141]. The set of points for which the dual leaves have maximal dimension is open and dense in M. This is in particular true for the dual foliation in (iii), the leaves of which are the boundary of the distance tubes of radius r around $S^2(1/2)$, $r \geq 0$. If the curvature assumption is dropped, then the dual foliation will not, in general, be metric. For example, any codimension one totally geodesic foliation $\mathcal{F}$ of hyperbolic space is the dual of the one-dimensional metric foliation orthogonal to it, but $\mathcal{F}$ is never metric in this situation.

(v) Let $\mathcal{F}$ be a metric foliation of a nonnegatively curved manifold, and suppose that the dual leaves are complete. Let $x \in M_p$ be a horizontal vector, $v \in M_p$ a vector orthogonal to the dual leaf through p. We claim that if c is the geodesic with $\dot{c}(0) = x$, and E the parallel field along c with $E(0) = v$, then the rectangle

$$\begin{aligned} V : \mathbb{R} \times [0,\infty) &\to M, \\ (t,s) &\mapsto \exp sE(t) \end{aligned}$$

is flat and totally geodesic. To see this, notice that the proof of Theorem 1.8.1 implies that E is a parallel Jacobi field which stays perpendicular to the dual leaf through p. By (iv), each geodesic $s \mapsto V_t(s) := \exp sE(t)$ is horizontal for the dual foliation, and is therefore vertical for $\mathcal{F}$. For any t, there exists $\epsilon > 0$ such that the restriction $c_{|[t,t+\epsilon]}$ is a local minimal connection between the leaves through $c(t)$ and $c(t+\epsilon)$. However, if $V_s(t) = V(t,s)$, then the restriction of V_s to $[t, t+\epsilon]$ also connects these leaves, and by the second Rauch comparison theorem, it cannot be longer than the restriction of c to the same interval, see Theorem 3.2.2. The equality version of that same theorem now implies the claim.

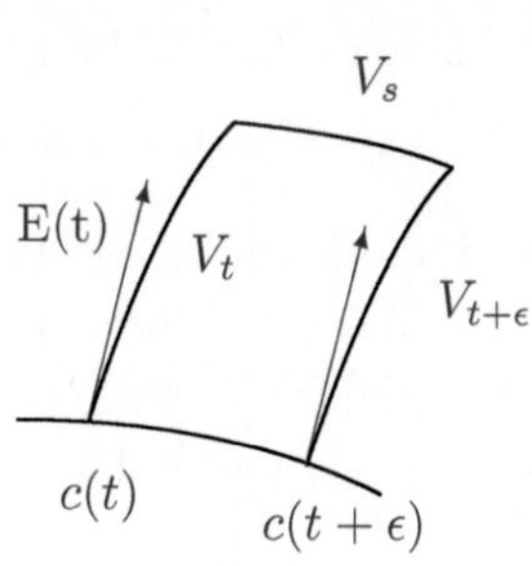

1.9 Basic identities

In this last section of the chapter, we gather for convenience of the reader a list of the definitions and identities introduced so far that are most often used throughout the text. The notation is the one agreed upon earlier, with X, Y denoting horizontal fields, T, U, V vertical ones, etc.

- A tensor:

$$A_X Y = \frac{1}{2}[X,Y]^{\mathbf{v}}, \quad \langle A_X^* U, Y\rangle = \langle A_X Y, U\rangle.$$

- S tensor and mean curvature form κ:

$$S_X U = -\nabla_U^{\mathbf{v}} X; \quad \kappa(E) = \operatorname{tr} S_{E^{\mathbf{h}}}; \quad d\kappa(X,Y) = -2\operatorname{div}(A_X Y) = \operatorname{tr} d^\nabla S_{X,Y}$$

(in the last equation, X and Y are assumed to be basic).

- Covariant derivative of a vector field E along a curve c:

$$\begin{aligned} E'^{\mathbf{h}} &= \widetilde{(\pi_* E)'} - A^*_{E^{\mathbf{h}}}\dot c^{\mathbf{v}} - A^*_{\dot c^{\mathbf{h}}} E^{\mathbf{v}} + \sigma(\dot c^{\mathbf{v}}, E^{\mathbf{v}}),\\ E'^{\mathbf{v}} &= A_{\dot c^{\mathbf{h}}} E^{\mathbf{h}} - S_{E^{\mathbf{h}}}\dot c^{\mathbf{v}} + E^{\mathbf{v}/\mathbf{v}}. \end{aligned}$$

- Curvature identities:

$$\begin{aligned} \pi_* R(x,y)z &= R^B(\pi_* x, \pi_* y,)\pi_* z + \pi_*(2A_z^* A_x y - A_x^* A_y z - A_y^* A_z x);\\ R^{\mathbf{v}}(x,y)z &= -(\nabla_z^{\mathbf{v}} A)_x y + S_z A_x y - S_y A_z x - S_x A_y z;\\ R^{\mathbf{v}}(x,u)y &= -(\nabla_u^{\mathbf{v}} A)_x y - A_y A_x^* u - (\nabla_x^{\mathbf{v}} S)_y u + S_y S_x u;\\ R^{\mathbf{v}}(x,y)u &= -2(\nabla_u^{\mathbf{v}} A)_x y - d^\nabla S_{x,y} u + [S_y, S_x]u + (A_x A_y^* - A_y A_x^*)u;\\ R^F(u,v)w &= R^{\mathbf{v}}(u,v)w + S_{\sigma(v,w)}u - S_{\sigma(u,w)}v;\\ R^{\mathbf{v}}(u,w)x &= (\nabla_w^{\mathbf{v}} S)_x u - (\nabla_u^{\mathbf{v}} S)_x w. \end{aligned}$$

- Sectional curvature identities:

$$\begin{aligned} K_{\pi_* x, \pi_* y} &= K_{x,y} + 3|A_x y|^2;\\ K^F_{u,v} &= K_{u,v} + \sigma(u,u)\sigma(v,v) - \sigma^2(u,v);\\ K_{x,u} &= \langle(\nabla_x^{\mathbf{v}} S)_x u, u\rangle + |A_x^* u|^2 - |S_x u|^2. \end{aligned}$$

- Holonomy Jacobi field J along a horizontal geodesic c:

$$J' = -A_{\dot c}^* J - S_{\dot c} J.$$

- Projectable Jacobi field J along a horizontal geodesic c:

$$J'^{\mathbf{v}} = -S_{\dot c} J^{\mathbf{v}} - A_{\dot c} J^{\mathbf{h}}.$$

- Riccati equation for S along a horizontal geodesic c:

$$R^{\mathbf{v}} = S_{\dot c}' - S_{\dot c}^2 + A_{\dot c} A_{\dot c}^*, \qquad R := R(\cdot, \dot c)\dot c.$$

Chapter 2

Basic Constructions and Examples

2.1 General vertical warping

Any Riemannian submersion can be used to generate new ones by deforming the metric in the vertical direction. To be specific, let $\pi : (M, \langle , \rangle) \to B$ be a Riemannian submersion. Given $\phi : M \to \mathbb{R}$, define a new metric $\langle , \rangle_\phi$ on M by

$$\langle e, f \rangle_\phi = e^{2\phi(p)} \langle e^{\mathbf{v}}, f^{\mathbf{v}} \rangle + \langle e^{\mathbf{h}}, f^{\mathbf{h}} \rangle, \qquad e, f \in M_p, \quad p \in M.$$

Since the horizontal metric is unchanged, $\pi : (M, \langle , \rangle_\phi) \to B$ is still a Riemannian submersion. X, Y, Z will denote basic fields, T_i vertical ones, and $\tilde{\nabla}$, $\tilde{R}$ the Levi-Civita connection and curvature tensor, respectively, of $\langle , \rangle_\phi$. We will assume that the deformation is constant along fibers, or equivalently, that the gradient of ϕ is basic.

2.1.1 The connection

We begin by computing the covariant derivatives. Since $[X, T]$ is vertical, we have:

$$\begin{aligned}\langle \tilde{\nabla}_T X, Y \rangle_\phi &= \langle \tilde{\nabla}_X T, Y \rangle_\phi = -\langle T, (\tilde{\nabla}_X Y)^{\mathbf{v}} \rangle_\phi = -\frac{1}{2} \langle T, [X, Y]^{\mathbf{v}} \rangle_\phi \\ &= -\frac{1}{2} e^{2\phi} \langle T, [X, Y]^{\mathbf{v}} \rangle = e^{2\phi} \langle \nabla_T X, Y \rangle,\end{aligned}$$

so that

$$(\tilde{\nabla}_T X)^{\mathbf{h}} = (\tilde{\nabla}_X T)^{\mathbf{h}} = e^{2\phi} (\nabla_T X)^{\mathbf{h}} = e^{2\phi} (\nabla_X T)^{\mathbf{h}}. \tag{2.1.1}$$

Similarly, using the formula for the Levi-Civita connection,

$$\begin{aligned}
2e^{2\phi}\langle\tilde{\nabla}_{T_1}X,T_2\rangle &= 2\langle\tilde{\nabla}_{T_1}X,T_2\rangle_\phi \\
&= X\langle T_1,T_2\rangle_\phi + \langle T_2,[T_1,X]\rangle_\phi - \langle T_1,[X,T_2]\rangle_\phi \\
&= 2X(\phi)e^{2\phi}\langle T_1,T_2\rangle + e^{2\phi}\{X\langle T_1,T_2\rangle + \langle T_2,[T_1,X]\rangle \\
&\qquad - \langle T_1,[X,T_2]\rangle\} \\
&= e^{2\phi}\{2\langle T_1,T_2\rangle\langle\nabla\phi,X\rangle + 2\langle\nabla_{T_1}X,T_2\rangle\}.
\end{aligned}$$

Thus,

$$(\tilde{\nabla}_T X)^{\mathbf{v}} = (\nabla_T X)^{\mathbf{v}} + \langle\nabla\phi,X\rangle T. \tag{2.1.2}$$

Much in the same way,

$$\begin{aligned}
2\langle\tilde{\nabla}_X T_1,T_2\rangle_\phi &= X\langle T_1,T_2\rangle_\phi + \langle T_2,[X,T_1]\rangle_\phi + \langle T_1,[T_2,X]\rangle_\phi \\
&= 2X(\phi)e^{2\phi}\langle T_1,T_2\rangle + e^{2\phi}\{X\langle T_1,T_2\rangle + \langle T_2,[X,T_1]\rangle \\
&\qquad + \langle T_1,[T_2,X]\rangle\} \\
&= e^{2\phi}\{2\langle T_1,T_2\rangle\langle\nabla\phi,X\rangle + 2\langle\nabla_X T_1,T_2\rangle\} \\
&= 2\langle\nabla\phi,X\rangle\langle T_1,T_2\rangle_\phi + 2\langle\nabla_X T_1,T_2\rangle_\phi,
\end{aligned}$$

which implies

$$(\tilde{\nabla}_X T)^{\mathbf{v}} = (\nabla_X T)^{\mathbf{v}} + \langle\nabla\phi,X\rangle. \tag{2.1.3}$$

For covariant derivatives of vertical fields in vertical directions, we compute:

$$\begin{aligned}
2\langle\tilde{\nabla}_{T_1}T_2,X\rangle = 2\langle\tilde{\nabla}_{T_1}T_2,X\rangle_\phi &= -X\langle T_2,T_1\rangle_\phi + \langle X,[T_2,T_1]\rangle_\phi + \langle T_2,[X,T_1]\rangle_\phi \\
&\quad - \langle T_1,[T_2,X]\rangle_\phi \\
&= -X(e^{2\phi})\langle T_2,T_1\rangle + e^{2\phi}(\langle X,[T_2,T_1]\rangle + \langle T_2,[X,T_1]\rangle \\
&\quad - \langle T_1,[T_2,X]\rangle) \\
&= -2e^{2\phi}\langle T_1,T_2\rangle\langle\nabla\phi,X\rangle + 2e^{2\phi}\langle\nabla_{T_1}T_2,X\rangle,
\end{aligned}$$

so that

$$(\tilde{\nabla}_{T_1}T_2)^{\mathbf{h}} = e^{2\phi}\{(\nabla_{T_1}T_2)^{\mathbf{h}} - \langle T_1,T_2\rangle\nabla\phi\}. \tag{2.1.4}$$

Next, we have

$$\begin{aligned}
2e^{2\phi}\langle\tilde{\nabla}_{T_1}T_2,T_3\rangle &= 2\langle\tilde{\nabla}_{T_1}T_2,T_3\rangle_\phi \\
&= T_1\langle T_2,T_3\rangle_\phi + T_2\langle T_3,T_1\rangle_\phi - T_3\langle T_1,T_2\rangle_\phi \\
&\qquad + \langle T_3,[T_1,T_2]\rangle_\phi + \langle T_2,[T_3,T_1]\rangle_\phi - \langle T_1,[T_2,T_3]\rangle_\phi \\
&= e^{2\phi}\{T_1\langle T_2,T_3\rangle + T_2\langle T_3,T_1\rangle - T_3\langle T_1,T_2\rangle \\
&\qquad + \langle T_3,[T_1,T_2]\rangle + \langle T_2,[T_3,T_1]\rangle - \langle T_1,[T_2,T_3]\rangle\} \\
&= 2e^{2\phi}\langle\nabla_{T_1}T_2,T_3\rangle,
\end{aligned}$$

so that

$$(\tilde{\nabla}_{T_1}T_2)^{\mathbf{v}} = (\nabla_{T_1}T_2)^{\mathbf{v}}. \tag{2.1.5}$$

The horizontal and vertical parts in the above identities may be combined into one. σ will denote the second fundamental tensor of the fibers in the original metric, $\sigma(T_1, T_2) = (\nabla_{T_1}T_2)^{\mathbf{h}}$, and similarly, the point-wise adjoint A^* of the A-tensor is the one with respect to the original metric. We then obtain:

$$\tilde{\nabla}_{T_1}T_2 = \nabla_{T_1}T_2 + (e^{2\phi} - 1)\sigma(T_1, T_2) - e^{2\phi}\langle T_1, T_2\rangle\nabla\phi, \tag{2.1.6}$$

$$\tilde{\nabla}_T X = \nabla_T X + (1 - e^{2\phi})A_X^* T + \langle\nabla\phi, X\rangle T, \tag{2.1.7}$$

$$\tilde{\nabla}_X T = \nabla_X T + (1 - e^{2\phi})A_X^* T + \langle\nabla\phi, X\rangle T. \tag{2.1.8}$$

It remains to compute $\tilde{\nabla}_X Y$. Since the metric is unchanged in the horizontal direction, $(\tilde{\nabla}_X Y)^{\mathbf{h}} = (\nabla_X Y)^{\mathbf{h}}$, whereas

$$2\langle\tilde{\nabla}_X Y, T\rangle_\phi = \langle[X,Y],T\rangle_\phi = e^{2\phi}\langle[X,Y],T\rangle = 2e^{2\phi}\langle\nabla_X Y, T\rangle = 2\langle\nabla_X Y, T\rangle_\phi.$$

Thus, $(\tilde{\nabla}_X Y)^{\mathbf{v}} = (\nabla_X Y)^{\mathbf{v}}$, and

$$\tilde{\nabla}_X Y = \nabla_X Y. \tag{2.1.9}$$

2.1.2 The curvature tensor

The curvature tensor $\tilde{R}$ of $\langle,\rangle_\phi$ can be readily derived from the above identities. Let us begin with terms involving horizontal vectors. In order to find the expression for $\tilde{R}(X,Y)Z$, we compute, using (2.1.9) and(2.1.8),

$$\begin{aligned}\tilde{\nabla}_X\tilde{\nabla}_Y Z &= \tilde{\nabla}_X\nabla_Y Z = \tilde{\nabla}_X((\nabla_Y Z)^{\mathbf{h}} + (\nabla_Y Z)^{\mathbf{v}}) \\ &= \nabla_X(\nabla_Y Z)^{\mathbf{h}} + \nabla_X(\nabla_Y Z)^{\mathbf{v}} + (1 - e^{2\phi})A_X^*(\nabla_Y Z)^{\mathbf{v}} \\ &\quad + \langle\nabla\phi, X\rangle(\nabla_Y Z)^{\mathbf{v}} \\ &= \nabla_X\nabla_Y Z + (1 - e^{2\phi})A_X^* A_Y Z + \langle\nabla\phi, X\rangle A_Y Z.\end{aligned}$$

Interchanging X and Y yields of course a corresponding expression for $\tilde{\nabla}_Y\tilde{\nabla}_X Z$. By (2.1.9) and (2.1.7),

$$\begin{aligned}\tilde{\nabla}_{[X,Y]}Z &= \tilde{\nabla}_{[X,Y]^{\mathbf{h}}}Z + \tilde{\nabla}_{[X,Y]^{\mathbf{v}}}Z \\ &= \nabla_{[X,Y]^{\mathbf{h}}}Z + \nabla_{[X,Y]^{\mathbf{v}}}Z + (1 - e^{2\phi})A_Z^*[X,Y]^{\mathbf{v}} + \langle\nabla\phi, Z\rangle[X,Y]^{\mathbf{v}} \\ &= \nabla_{[X,Y]}Z + 2(1 - e^{2\phi})A_Z^* A_X Y + 2\langle\nabla\phi, Z\rangle A_X Y.\end{aligned}$$

Adding these identities then yields

$$\begin{aligned}\tilde{R}(X,Y)Z = R(X,Y)Z &+ (1 - e^{2\phi})(A_X^* A_Y Z - A_Y^* A_X Z - 2A_Z^* A_X Y) \\ &+ \langle\nabla\phi, X\rangle A_Y Z - \langle\nabla\phi, Y\rangle A_X Z - 2\langle\nabla\phi, Z\rangle A_X Y.\end{aligned} \tag{2.1.10}$$

Taking vertical and horizontal parts, we obtain

$$\tilde{R}^{\mathbf{v}}(X,Y)Z = R^{\mathbf{v}}(X,Y)Z + \langle\nabla\phi, X\rangle A_Y Z - \langle\nabla\phi, Y\rangle A_X Z - 2\langle\nabla\phi, Z\rangle A_X Y, \tag{2.1.11}$$

and

$$\tilde{R}^{\mathbf{h}}(X,Y)Z = R^{\mathbf{h}}(X,Y)Z + (1-e^{2\phi})(A_X^* A_Y Z - A_Y^* A_X Z - 2A_Z^* A_X Y). \tag{2.1.12}$$

Recalling the relation between the curvatures R of M and R_B of B, the last identity may be rewritten as

$$\tilde{R}^{\mathbf{h}}(X,Y)Z = (1-e^{2\phi})R_B(X,Y)Z + e^{2\phi}R^{\mathbf{h}}(X,Y)Z, \tag{2.1.13}$$

where we denote by the same letter a basic field on M and the vector field on B that is π-related to it.

Next, we look at terms that only involve vertical vectors.

To find $\tilde{R}^{\mathbf{h}}(T_1,T_2)T_3$, we first calculate $(\tilde{\nabla}_T X)^{\mathbf{h}}$ for nonbasic X:

$$\begin{aligned}
(\tilde{\nabla}_T X)^{\mathbf{h}} &= (\tilde{\nabla}_X T)^{\mathbf{h}} + [T,X]^{\mathbf{h}} = e^{2\phi}(\nabla_X T)^{\mathbf{h}} + [T,X]^{\mathbf{h}} \\
&= e^{2\phi}\{(\nabla_X T)^{\mathbf{h}} + [T,X]^{\mathbf{h}}\} + (1-e^{2\phi})[T,X]^{\mathbf{h}} \\
&= e^{2\phi}\nabla_X T + (1-e^{2\phi})[T,X]^{\mathbf{h}}.
\end{aligned}$$

Using (2.1.6), we then obtain

$$\begin{aligned}
\big(\tilde{\nabla}_{T_1}(\tilde{\nabla}_{T_2}T_3)^{\mathbf{h}}\big)^{\mathbf{h}} &= \big(\tilde{\nabla}_{T_1} e^{2\phi}\{(\nabla_{T_2}T_3)^{\mathbf{h}} - \langle T_2,T_3\rangle\nabla\phi\}\big)^{\mathbf{h}} \\
&= e^{2\phi}\big\{\big(\tilde{\nabla}_{T_1}(\nabla_{T_2}T_3)^{\mathbf{h}}\big)^{\mathbf{h}} - T_1\langle T_2,T_3\rangle\nabla\phi - \langle T_2,T_3\rangle\tilde{\nabla}_{T_1}\nabla\phi\big\}^{\mathbf{h}} \\
&= e^{2\phi}\big\{\big(\nabla_{T_1}(\nabla_{T_2}T_3)^{\mathbf{h}}\big)^{\mathbf{h}} + (1-e^{2\phi})\big([T_1,(\nabla_{T_2}T_3)^{\mathbf{h}}]^{\mathbf{h}} \\
&\quad - (\nabla_{\sigma(T_2,T_3)}T_1)^{\mathbf{h}}\big) - T_1\langle T_2,T_3\rangle\nabla\phi - \langle T_2,T_3\rangle(\nabla_{T_1}\nabla\phi)^{\mathbf{h}} \\
&\quad - \langle T_2,T_3\rangle(1-e^{2\phi})A^*_{\nabla\phi}T_1\big\} \\
&= e^{2\phi}\big\{\big(\nabla_{T_1}(\nabla_{T_2}T_3)^{\mathbf{h}}\big)^{\mathbf{h}} + (1-e^{2\phi})\big(\nabla_{T_1}(\nabla_{T_2}T_3)^{\mathbf{h}}\big)^{\mathbf{h}} \\
&\quad - T_1\langle T_2,T_3\rangle\nabla\phi - \langle T_2,T_3\rangle(\nabla_{T_1}\nabla\phi)^{\mathbf{h}} \\
&\quad - \langle T_2,T_3\rangle(1-e^{2\phi})A^*_{\nabla\phi}T_1\big\} \\
&= e^{2\phi}\big\{(2-e^{2\phi})\big(\nabla_{T_1}(\nabla_{T_2}T_3)^{\mathbf{h}}\big)^{\mathbf{h}} - T_1\langle T_2,T_3\rangle\nabla\phi \\
&\quad + \langle T_2,T_3\rangle e^{2\phi}A^*_{\nabla\phi}T_1\big\},
\end{aligned}$$

and

$$\begin{aligned}
\big(\tilde{\nabla}_{T_1}(\tilde{\nabla}_{T_2}T_3)^{\mathbf{v}}\big)^{\mathbf{h}} = \big(\tilde{\nabla}_{T_1}(\nabla_{T_2}T_3)^{\mathbf{v}}\big)^{\mathbf{h}} = e^{2\phi}\big\{&\big(\nabla_{T_1}(\nabla_{T_2}T_3)^{\mathbf{v}}\big)^{\mathbf{h}} \\
&- \langle T_1,\nabla_{T_2}T_3\rangle\nabla\phi\big\}.
\end{aligned}$$

Assuming without loss of generality that $[T_1, T_2] = 0$,

$$\begin{aligned}\tilde{R}^{\mathbf{h}}(T_1,T_2)T_3 = e^{2\phi}\{&(2-e^{2\phi}))R^{\mathbf{h}}(T_1,T_2)T_3 + \big(-T_1\langle T_2,T_3\rangle + T_2\langle T_1,T_3\rangle \\ &- \langle T_1, \nabla_{T_2}T_3\rangle + \langle T_2, \nabla_{T_1}T_3\rangle\big)\nabla\phi \\ &+ e^{2\phi}A^*_{\nabla\phi}\big(\langle T_2,T_3\rangle T_1 - \langle T_1,T_3\rangle T_2\big)\big\}.\end{aligned}$$

The coefficient of $\nabla\phi$ in the above identity is zero, however, so that we finally obtain:

$$\begin{aligned}e^{-2\phi}\tilde{R}^{\mathbf{h}}(T_1,T_2)T_3 = {}&(2-e^{2\phi})R^{\mathbf{h}}(T_1,T_2)T_3 \\ &+ e^{2\phi}A^*_{\nabla\phi}\big(\langle T_2,T_3\rangle T_1 - \langle T_1,T_3\rangle T_2\big).\end{aligned} \tag{2.1.14}$$

In order to express the vertical component of $\tilde{R}(T_1,T_2)T_3$, we use (2.1.6) and (2.1.7):

$$\big(\tilde{\nabla}_{T_1}(\tilde{\nabla}_{T_2}T_3)^{\mathbf{v}}\big)^{\mathbf{v}} = \big(\tilde{\nabla}_{T_1}(\nabla_{T_2}T_3)^{\mathbf{v}}\big)^{\mathbf{v}} = \big(\nabla_{T_1}(\nabla_{T_2}T_3)^{\mathbf{v}}\big)^{\mathbf{v}},$$

whereas

$$\begin{aligned}\big(\tilde{\nabla}_{T_1}(\tilde{\nabla}_{T_2}T_3)^{\mathbf{h}}\big)^{\mathbf{v}} &= \big(\tilde{\nabla}_{T_1}(e^{2\phi}\{(\nabla_{T_2}T_3)^{\mathbf{h}} - \langle T_2,T_3\rangle\nabla\phi\})\big)^{\mathbf{v}} \\ &= e^{2\phi}\big\{\nabla_{T_1}\big((\nabla_{T_2}T_3)^{\mathbf{h}} - \langle T_2,T_3\rangle\nabla\phi\big)^{\mathbf{v}} \\ &\qquad + \langle\nabla\phi, (\nabla_{T_2}T_3)^{\mathbf{h}} - \langle T_2,T_3\rangle\nabla\phi\rangle T_1\big\} \\ &= e^{2\phi}\big\{\nabla_{T_1}\big((\nabla_{T_2}T_3)^{\mathbf{h}}\big)^{\mathbf{v}} + \langle T_2,T_3\rangle\big(S_{\nabla\phi}T_1 - |\nabla\phi|^2T_1\big) \\ &\qquad + \langle\nabla\phi, \nabla_{T_2}T_3\rangle T_1\big\} \\ &= \nabla_{T_1}\big((\nabla_{T_2}T_3)^{\mathbf{h}}\big)^{\mathbf{v}} + (1-e^{2\phi})S_{\sigma(T_2,T_3)}T_1 \\ &\qquad + e^{2\phi}\big\{\langle T_2,T_3\rangle\big(S_{\nabla\phi}T_1 - |\nabla\phi|^2T_1\big) \\ &\qquad + \langle\nabla\phi, \sigma(T_2,T_3)\rangle T_1\big\}.\end{aligned}$$

Thus,

$$\begin{aligned}\tilde{R}^{\mathbf{v}}(T_1,T_2)T_3 = R^{\mathbf{v}}(T_1,T_2)T_3 &+ (1-e^{2\phi})\{S_{\sigma(T_2,T_3)}T_1 - S_{\sigma(T_1,T_3)}T_2\} \\ &+ e^{2\phi}\big\{(S_{\nabla\phi} - |\nabla\phi|^2I)(\langle T_2,T_3\rangle T_1 - \langle T_1,T_3\rangle T_2) \\ &\qquad + \langle\nabla\phi,\sigma(T_2,T_3)\rangle T_1 - \langle\nabla\phi,\sigma(T_1,T_3)\rangle T_2\big\}.\end{aligned} \tag{2.1.15}$$

This identity may be rewritten as follows: According to the Gauss equation,

$$R^{\mathbf{v}}(T_1,T_2)T_3 = R^F(T_1,T_2)T_3 + S_{\sigma(T_1,T_3)}T_2 - S_{\sigma(T_2,T_3)}T_1,$$

where R^F denotes the intrinsic curvature of the fiber (with respect to the original metric). Substituting this into (2.1.15) then yields

$$\begin{aligned}\tilde{R}^{\mathbf{v}}(T_1,T_2)T_3 = (1-e^{2\phi})&R^F(T_1,T_2)T_3 + e^{2\phi}R^{\mathbf{v}}(T_1,T_2)T_3 \\ &+ e^{2\phi}\big\{(S_{\nabla\phi} - |\nabla\phi|^2I)(\langle T_2,T_3\rangle T_1 - \langle T_1,T_3\rangle T_2)\big\} \\ &+ e^{2\phi}\big\{\langle\nabla\phi,\sigma(T_2,T_3)\rangle T_1 - \langle\nabla\phi,\sigma(T_1,T_3)\rangle T_2\big\}.\end{aligned} \tag{2.1.16}$$

We next compute $\tilde{R}^{\mathbf{v}}(X,T)Y$, which involves the following five expressions:

$$\begin{aligned}(\tilde{\nabla}_X(\tilde{\nabla}_T Y)^{\mathbf{v}})^{\mathbf{v}} &= (\tilde{\nabla}_X\{(\nabla_T Y)^{\mathbf{v}} + \langle\nabla\phi, Y\rangle T\})^{\mathbf{v}} \\ &= (\nabla_X\{(\nabla_T Y)^{\mathbf{v}} + \langle\nabla\phi, Y\rangle T\})^{\mathbf{v}} + \langle\nabla\phi, X\rangle((\nabla_T Y)^{\mathbf{v}} \\ &\qquad + \langle\nabla\phi, Y\rangle T) \\ &= (\nabla_X(\nabla_T Y)^{\mathbf{v}})^{\mathbf{v}} + \{\langle\nabla_X\nabla\phi, Y\rangle + \langle\nabla\phi, \nabla_X Y\rangle \\ &\qquad + \langle\nabla\phi, X\rangle\langle\nabla\phi, Y\rangle\}T + \langle\nabla\phi, Y\rangle(\nabla_X T)^{\mathbf{v}} \\ &\qquad + \langle\nabla\phi, X\rangle(\nabla_T Y)^{\mathbf{v}},\end{aligned}$$

$$\begin{aligned}(\tilde{\nabla}_X(\tilde{\nabla}_T Y)^{\mathbf{h}})^{\mathbf{v}} &= (\tilde{\nabla}_X(e^{2\phi}\nabla_T Y)^{\mathbf{h}})^{\mathbf{v}} = e^{2\phi}(\nabla_X(\nabla_T Y)^{\mathbf{h}})^{\mathbf{v}}, \\ (\tilde{\nabla}_T(\tilde{\nabla}_X Y)^{\mathbf{h}})^{\mathbf{v}} &= (\tilde{\nabla}_T(\nabla_X Y)^{\mathbf{h}})^{\mathbf{v}} = (\nabla_T(\nabla_X Y)^{\mathbf{h}})^{\mathbf{v}} + \langle\nabla\phi, \nabla_X Y\rangle T, \\ (\tilde{\nabla}_T(\tilde{\nabla}_X Y)^{\mathbf{v}})^{\mathbf{v}} &= (\tilde{\nabla}_T(\nabla_X Y)^{\mathbf{v}})^{\mathbf{v}} = (\nabla_T(\nabla_X Y)^{\mathbf{v}})^{\mathbf{v}}, \\ (\tilde{\nabla}_{[X,T]}Y)^{\mathbf{v}} &= (\nabla_{[X,T]}Y)^{\mathbf{v}} + \langle\nabla\phi, Y\rangle[X,T].\end{aligned}$$

Let h_ϕ denote the Hessian form of ϕ, $h_\phi(X,Y) = \langle\nabla_X\nabla\phi, Y\rangle = \langle X, \nabla_Y\nabla\phi\rangle$. Adding the above expressions then yields

$$\begin{aligned}\tilde{R}^{\mathbf{v}}(X,T)Y = R^{\mathbf{v}}(X,T)Y &- (1-e^{2\phi})(\nabla_X(\nabla_T Y)^{\mathbf{h}})^{\mathbf{v}} + \{h_\phi(X,Y) \\ &+ \langle\nabla\phi, X\rangle\langle\nabla\phi, Y\rangle\}T + \langle\nabla\phi, Y\rangle(\nabla_T X)^{\mathbf{v}} + \langle\nabla\phi, X\rangle(\nabla_T Y)^{\mathbf{v}}.\end{aligned}$$

Equivalently,

$$\begin{aligned}\tilde{R}^{\mathbf{v}}(X,T)Y = R^{\mathbf{v}}(X,T)Y &+ (1-e^{2\phi})A_X A_Y^* T + \{h_\phi(X,Y) \\ &+ \langle\nabla\phi, X\rangle\langle\nabla\phi, Y\rangle\}T - (\langle\nabla\phi, X\rangle S_Y T + \langle\nabla\phi, Y\rangle S_X T).\end{aligned} \tag{2.1.17}$$

All other curvature identities can be derived from the previous ones using symmetries of the curvature tensor. For example, to obtain $\tilde{R}^{\mathbf{h}}(X,T)Y$, (2.1.11) implies

$$\begin{aligned}\langle\tilde{R}^{\mathbf{h}}(X,T)Y, Z\rangle &= \langle\tilde{R}^{\mathbf{h}}(X,T)Y, Z\rangle_\phi = \langle\tilde{R}^{\mathbf{v}}(Y,Z)X, T\rangle_\phi \\ &= \langle R^{\mathbf{v}}(Y,Z)X, T\rangle_\phi + \langle\nabla\phi, Y\rangle\langle A_Z X, T\rangle_\phi \\ &\qquad - \langle\nabla\phi, Z\rangle\langle A_Y X, T\rangle_\phi - 2\langle\nabla\phi, X\rangle\langle A_Y Z, T\rangle_\phi \\ &= e^{2\phi}\{\langle R^{\mathbf{v}}(Y,Z)X, T\rangle + \langle\nabla\phi, Y\rangle\langle A_Z X, T\rangle \\ &\qquad - \langle\nabla\phi, Z\rangle\langle A_Y X, T\rangle - 2\langle\nabla\phi, X\rangle\langle A_Y Z, T\rangle\} \\ &= e^{2\phi}\{\langle R^{\mathbf{h}}(X,T)Y, Z\rangle - \langle\nabla\phi, Y\rangle\langle A_X^* T, Z\rangle \\ &\qquad - \langle\nabla\phi, Z\rangle\langle A_Y X, T\rangle - 2\langle\nabla\phi, X\rangle\langle A_Y^* T, Z\rangle\},\end{aligned}$$

so that

$$e^{-2\phi}\tilde{R}^{\mathbf{h}}(X,T)Y = R^{\mathbf{h}}(X,T)Y - \langle\nabla\phi, Y\rangle A_X^* T - 2\langle\nabla\phi, X\rangle A_Y^* T + \langle A_X Y, T\rangle\nabla\phi. \tag{2.1.18}$$

Similarly, to compute $\tilde{R}^{\mathbf{h}}(T_1,X)T_2$, one can proceed as follows:

$$\begin{aligned}
\langle \tilde{R}^{\mathbf{h}}(T_1,X)T_2,Y\rangle &= \langle \tilde{R}^{\mathbf{h}}(T_1,X)T_2,Y\rangle_\phi = \langle \tilde{R}^{\mathbf{v}}(X,T_1)Y,T_2\rangle_\phi \\
&= \langle R(X,T_1)Y,T_2\rangle_\phi + (1-e^{2\phi})\langle A_X A_Y^* T_1,T_2\rangle_\phi \\
&\quad + \{h_\phi(X,Y) + \langle\nabla\phi,X\rangle\langle\nabla\phi,Y\rangle\}\langle T_1,T_2\rangle_\phi \\
&\quad - \langle\nabla\phi,X\rangle\langle S_Y T_1,T_2\rangle_\phi - \langle\nabla\phi,Y\rangle\langle S_X T_1,T_2\rangle_\phi \\
&= \mathbf{e}^{2\phi}\{\langle R(X,T_1)Y,T_2\rangle + (1-e^{2\phi})\langle A_X A_Y^* T_1,T_2\rangle \\
&\quad + \{h_\phi(X,Y) + \langle\nabla\phi,X\rangle\langle\nabla\phi,Y\rangle\}\langle T_1,T_2\rangle \\
&\quad - \langle\nabla\phi,X\rangle\langle S_Y T_1,T_2\rangle - \langle\nabla\phi,Y\rangle\langle S_X T_1,T_2\rangle\} \\
&= \mathbf{e}^{2\phi}\{\langle R(T_1,X)T_2,Y\rangle - (1-e^{2\phi})\langle A^*_{A^*_X T_2}T_1,Y\rangle \\
&\quad + \langle T_1,T_2\rangle\{\langle\nabla_X\nabla\phi,Y\rangle + \langle\nabla\phi,X\rangle\langle\nabla\phi,Y\rangle\} \\
&\quad - \langle\nabla\phi,X\rangle\langle\nabla_{T_1}T_2,Y\rangle - \langle S_X T_1,T_2\rangle\langle\nabla\phi,Y\rangle\}.
\end{aligned}$$

Thus,

$$\begin{aligned}
e^{-2\phi}\tilde{R}^{\mathbf{h}}(T_1,X)T_2 = R^{\mathbf{h}}(T_1,X)T_2 &- (1-e^{2\phi})A^*_{A^*_X T_2}T_1 + \langle T_1,T_2\rangle\langle\nabla\phi,X\rangle\nabla\phi \\
&- \langle X,\sigma(T_1,T_2)\rangle\nabla\phi - \langle\nabla\phi,X\rangle\sigma(T_1,T_2) \\
&+ \langle T_1,T_2\rangle(\nabla_X\nabla\phi)^{\mathbf{h}}.
\end{aligned} \tag{2.1.19}$$

For $\tilde{R}^{\mathbf{h}}(T_1,T_2)X$, we use the Bianchi identity and (2.1.19):

$$\begin{aligned}
e^{-2\phi}\tilde{R}^{\mathbf{h}}(T_1,T_2)X &= e^{-2\phi}(\tilde{R}^{\mathbf{h}}(T_1,X)T_2) + \tilde{R}^{\mathbf{h}}(X,T_2)T_1) \\
&= R^{\mathbf{h}}(T_1,X)T_2 + R^{\mathbf{h}}(X,T_2)T_1 - (1-e^{2\phi})A^*_{A^*_X T_2}T_1 \\
&\quad + (1-e^{2\phi})A^*_{A^*_X T_1}T_2,
\end{aligned}$$

so that

$$e^{-2\phi}\tilde{R}^{\mathbf{h}}(T_1,T_2)X = R^{\mathbf{h}}(T_1,T_2)X + (1-e^{2\phi})(A^*_{A^*_X T_1}T_2 - A^*_{A^*_X T_2}T_1). \tag{2.1.20}$$

The latter identity can also be used to derive $\tilde{R}^{\mathbf{v}}(X,Y)T$: Notice that

$$\langle A^*_{A^*_X T}T_1,Y\rangle = \langle T_1, A_{A^*_X T}Y\rangle = -\langle T_1, A_Y A^*_X T\rangle.$$

Thus,

$$\begin{aligned}
\langle \tilde{R}(X,Y)T,T_1\rangle &= e^{-2\phi}\langle \tilde{R}(X,Y)T,T_1\rangle_\phi = e^{-2\phi}\langle \tilde{R}(T,T_1)X,Y\rangle_\phi \\
&= e^{-2\phi}\langle \tilde{R}(T,T_1)X,Y\rangle \\
&= \langle R(T,T_1)X,Y\rangle + (1-e^{2\phi})(-\langle A_Y A^*_X T,T_1\rangle \\
&\quad + \langle T, A_Y A^*_X T_1\rangle) \\
&= \langle R(X,Y)T,T_1\rangle + (1-e^{2\phi})\langle (A_X A^*_Y - A_Y A^*_X)T,T_1\rangle,
\end{aligned}$$

and

$$\tilde{R}^{\mathbf{v}}(X,Y)T = R^{\mathbf{v}}(X,Y)T + (1-e^{2\phi})(A_X A_Y^* - A_Y A_X^*)T. \tag{2.1.21}$$

Finally, by (2.1.11),

$$\begin{aligned}
\langle \tilde{R}(X,Y)T, Z\rangle &= \langle \tilde{R}(X,Y)T, Z\rangle_\phi = -\langle \tilde{R}(X,Y)Z, T\rangle_\phi \\
&= -e^{2\phi}\langle \tilde{R}(X,Y)Z, T\rangle \\
&= -e^{2\phi}\{\langle R(X,Y)Z,T\rangle + \langle \nabla\phi, X\rangle\langle A_Y Z, T\rangle \\
&\qquad - \langle \nabla\phi, Y\rangle\langle A_X Z, T\rangle - 2\langle \nabla\phi, Z\rangle\langle A_X Y, T\rangle\} \\
&= e^{2\phi}\{\langle R(X,Y)T,Z\rangle - \langle \nabla\phi, X\rangle\langle A_Y^* T, Z\rangle \\
&\qquad + \langle \nabla\phi, Y\rangle\langle A_X^* T, Z\rangle + 2\langle \nabla\phi, Z\rangle\langle A_X Y, T\rangle\},
\end{aligned}$$

so that

$$\begin{aligned}
e^{-2\phi}\tilde{R}^{\mathbf{h}}(X,Y)T = R^{\mathbf{h}}(X,Y)T &- \langle \nabla\phi, X\rangle A_Y^* T + \langle \nabla\phi, Y\rangle A_X^* T \\
&+ 2\langle A_X Y, T\rangle \nabla\phi.
\end{aligned} \tag{2.1.22}$$

2.1.3 The sectional curvatures

The results from the previous section yield corresponding identities for the sectional curvatures. Given (not necessarily orthonormal) vertical v, w and horizontal x, y in M_p, $p \in M$, let

$$\begin{aligned}
k(v,w) &:= \langle R(v,w)w, v\rangle, \\
\tilde{k}(v,w) &:= \langle R(v,w)w, v\rangle_\phi, \\
k_B(x,y) &:= \langle R_B(\pi_* x, \pi_* y)\pi_* y, \pi_* x\rangle_B,
\end{aligned}$$

etc. Then

$$\tilde{k}(X,Y) = (1-e^{2\phi})k_B(X,Y) + e^{2\phi}k(X,Y), \tag{2.1.23}$$

$$\begin{aligned}
\tilde{k}(X,T) = k(X,T) &- (1-e^{2\phi})|A_X^* T|^2 + 2\langle \nabla\phi, X\rangle\langle X, \sigma(T,T)\rangle \\
&- (h_\phi(X,X) + \langle \nabla\phi, X\rangle^2)|T|^2,
\end{aligned} \tag{2.1.24}$$

$$\begin{aligned}
e^{-4\phi}\tilde{k}(T_1,T_2) = e^{-2\phi}&(1-e^{2\phi})k_F(T_1,T_2) + k(T_1,T_2) \\
&+ \langle \nabla\phi, \sigma(T_1,T_1)\rangle|T_2|^2 + \langle \nabla\phi, \sigma(T_2,T_2)\rangle|T_1|^2 \\
&- 2\langle \nabla\phi, \sigma(T_1,T_2)\rangle\langle T_1,T_2\rangle - |\nabla\phi|^2\big(|T_1|^2|T_2|^2 \\
&- \langle T_1,T_2\rangle^2\big).
\end{aligned} \tag{2.1.25}$$

2.1.4 The Ricci curvature

Let k denote the dimension of the fibers. Recall from Definition 1.4.4 that the *mean curvature form* of the submersion is the horizontal 1-form κ, where $\kappa(u) = \operatorname{tr} S_{u^{\mathbf{h}}}$, and the *mean curvature vector field* is the (horizontal) vector field H dual to κ: $\langle H, Y\rangle = \kappa(Y)$. If T_i is a local orthonormal basis of vertical vector fields, and X_j a local basic one, then locally, $H = \sum_i \nabla_{T_i} T_i$, because

$$\langle H, Y\rangle = \operatorname{tr} S_Y = \sum_i \langle S_Y T_i, T_i\rangle = -\sum_i \langle \nabla_{T_i} Y, T_i\rangle = \Big\langle \sum_i \nabla_{T_i} T_i \Big\rangle Y.$$

The Laplacians $\Delta_B\phi$ and $\Delta\phi$ of ϕ on B and M are related by $\Delta_B\phi = \Delta\,\phi + \kappa(\nabla\phi)$, since

$$\begin{aligned}\Delta_B\phi &= \sum_j h_\phi(X_j, X_j) = \Delta\,\phi - \sum_i h_\phi(T_i, T_i) = \Delta\,\phi - \sum_i \langle \nabla_{T_i}\nabla\phi, T_i\rangle \\ &= \Delta\,\phi + \langle \nabla\phi, H\rangle.\end{aligned}$$

We denote by $\operatorname{Ric}^{\mathbf{h}}(u,v)$ the trace of the operator $w \mapsto R(w,u)v$ restricted to the horizontal space, and by $\operatorname{Ric}^{\mathbf{v}}$ the one restricted to the vertical space. As usual, a basic field on M and the π-related field on B are denoted by the same letter. Let $\tilde{T}_j := e^{-\phi} T_j$, so that $\tilde{T}_j$ is a local orthonormal basis of the vertical space in the ϕ-metric. Now,

$$\begin{aligned}\widetilde{\operatorname{Ric}}(X,Y) &= \widetilde{\operatorname{Ric}}^{\mathbf{h}}(X,Y) + \widetilde{\operatorname{Ric}}^{\mathbf{v}}(X,Y) \\ &= \sum_{i=1}^{n} \langle \tilde{R}(X_i, X)Y, X_i\rangle_\phi + \sum_{j=1}^{k} \langle \tilde{R}(\tilde{T}_j, X)Y, \tilde{T}_j\rangle_\phi,\end{aligned}$$

where

$$\widetilde{\operatorname{Ric}}^{\mathbf{h}}(X,Y) = (1 - e^{2\phi}) \operatorname{Ric}_B(X,Y) + e^{2\phi} \operatorname{Ric}^{\mathbf{h}}(X,Y) \tag{2.1.26}$$

by (2.1.13). On the other hand, (2.1.17) implies

$$\begin{aligned}\widetilde{\operatorname{Ric}}^{\mathbf{v}}(X,Y) &= \sum_j \langle \tilde{R}((\tilde{T}_j, X)Y, \tilde{T}_j\rangle_\phi = e^{2\phi} \sum_j \langle \tilde{R}(\tilde{T}_j, X)Y, \tilde{T}_j\rangle \\ &= \sum_j \langle \tilde{R}(T_j, X)Y, T_j\rangle = -\sum_j \langle \tilde{R}(X, T_j)Y, T_j\rangle \\ &= -\sum_j \langle R(X, T_j)Y, T_j\rangle - (1 - e^{2\phi}) \sum_j \langle A_X A_Y^* T_j, T_j\rangle \\ &\quad - \sum_j \{h_\phi(X,Y) + \langle \nabla\phi, X\rangle\langle \nabla\phi, Y\rangle\}\langle T_j, T_j\rangle \\ &\quad + \langle \nabla\phi, X\rangle \sum_j \langle S_Y T_j, T_j\rangle + \langle \nabla\phi, Y\rangle \sum_j \langle S_X T_j, T_j\rangle \\ &= \operatorname{Ric}^{\mathbf{v}}(X,Y) - (1 - e^{2\phi}) \operatorname{tr} A_X A_Y^* - k\{h_\phi(X,Y) \\ &\quad + \langle \nabla\phi, X\rangle\langle \nabla\phi, Y\rangle\} + \langle \nabla\phi, X\rangle\langle Y, H\rangle + \langle \nabla\phi, Y\rangle\langle X, H\rangle.\end{aligned}$$

Recalling that $\operatorname{tr} A_X A_Y^* = (1/3)(\mathrm{Ric}_B(X,Y) - \mathrm{Ric}^{\mathbf{h}}(X,Y))$, we obtain

$$\begin{aligned}\widetilde{\mathrm{Ric}}^{\mathbf{v}}(X,Y) = \mathrm{Ric}^{\mathbf{v}}(X,Y) &- \frac{1}{3}(1-e^{2\phi})\big(\mathrm{Ric}_B(X,Y) - \mathrm{Ric}^{\mathbf{h}}(X,Y)\big)\\ &- k\{h_\phi(X,Y) + \langle\nabla\phi, X\rangle\langle\nabla\phi, Y\rangle\} + \langle\nabla\phi, X\rangle\langle Y, H\rangle\\ &+ \langle\nabla\phi, Y\rangle\langle X, H\rangle.\end{aligned} \tag{2.1.27}$$

Adding (2.1.26) and (2.1.27) then yields

$$\begin{aligned}\widetilde{\mathrm{Ric}}(X,Y) = \mathrm{Ric}(X,Y) &+ \frac{2}{3}(1-e^{2\phi})\big(\mathrm{Ric}_B(X,Y) - \mathrm{Ric}^{\mathbf{h}}(X,Y)\big)\\ &+ \langle\nabla\phi, X\rangle\langle Y, H\rangle + \langle\nabla\phi, Y\rangle\langle X, H\rangle - k\{h_\phi(X,Y)\\ &+ \langle\nabla\phi, X\rangle\langle\nabla\phi, Y\rangle\}.\end{aligned} \tag{2.1.28}$$

For the "vertizontal" Ricci curvature $\widetilde{\mathrm{Ric}}(X,T)$, we have

$$\begin{aligned}\widetilde{\mathrm{Ric}}(X,T) &= \widetilde{\mathrm{Ric}}^{\mathbf{h}}(X,T) + \widetilde{\mathrm{Ric}}^{\mathbf{v}}(X,T)\\ &= \sum_i \langle\tilde{R}(X_i, X)T, X_i\rangle_\phi + \sum_j \langle\tilde{R}(\tilde{T}_j, X)T, \tilde{T}_j\rangle_\phi,\end{aligned}$$

where, by (2.1.11),

$$\begin{aligned}\langle\tilde{R}(X_i, X)T, X_i\rangle_\phi &= -\langle\tilde{R}(X_i, X)X_i, T\rangle_\phi = -e^{2\phi}\langle\tilde{R}(X_i, X)X_i, T\rangle\\ &= -e^{2\phi}\{\langle R(X_i, X)X_i, T\rangle + \langle\nabla\phi, X_i\rangle\langle A_X X_i, T\rangle\\ &\quad - 2\langle\nabla\phi, X_i\rangle\langle A_{X_i}X, T\rangle\}\\ &= e^{2\phi}\{\langle R(X_i, X)T, X\rangle_i + 3\langle\nabla\phi, X_i\rangle\langle A_{X_i}X, T\rangle\}.\end{aligned}$$

Thus,

$$\widetilde{\mathrm{Ric}}^{\mathbf{h}}(X,T) = e^{2\phi}\{\mathrm{Ric}^{\mathbf{h}}(X,T) + 3\langle A_{\nabla\phi}X, T\rangle\}. \tag{2.1.29}$$

Furthermore, by (2.1.14),

$$\begin{aligned}\langle\tilde{R}(\tilde{T}_j, X)T, \tilde{T}_j\rangle_\phi &= \langle\tilde{R}(T, \tilde{T}_j)\tilde{T}_j, X\rangle_\phi = e^{-2\phi}\langle\tilde{R}(T, T_j)T_j, X\rangle\\ &= \langle R(T, T_j)T_j, X\rangle + (1-e^{2\phi})\langle A^*_{\sigma(T_j,T_j)}T - A^*_{\sigma(T,T_j)}T_j, X\rangle\\ &\quad + e^{2\phi}\langle A^*_{\nabla\phi}(T - \langle T, T_j\rangle T_j), X\rangle,\end{aligned}$$

so that

$$\begin{aligned}\widetilde{\mathrm{Ric}}^{\mathbf{v}}(X,T) = \mathrm{Ric}^{\mathbf{v}}(X,T) &+ e^{2\phi}(k-1)\langle A_{\nabla\phi}X, T\rangle\\ &+ (1-e^{2\phi})\{\mathrm{tr}(U \mapsto A_X\sigma(T,U)) - \langle A_X^*T, H\rangle\}.\end{aligned} \tag{2.1.30}$$

Adding (2.1.29) and (2.1.30), we obtain

$$\begin{aligned}\widetilde{\mathrm{Ric}}(X,T) = \mathrm{Ric}^{\mathbf{v}}(X,T) &+ e^{2\phi}\big(\mathrm{Ric}^{\mathbf{h}}(X,T) + (k+2)\langle A_{\nabla\phi}X, T\rangle\big)\\ &+ (1-e^{2\phi})\big(\operatorname{tr} A_X\sigma(T,\cdot) - \langle A_X^*T, H\rangle\big).\end{aligned} \tag{2.1.31}$$

In order to derive the vertical Ricci curvature, we temporarily introduce vertical fields U and V, since the notation T_i is already reserved for a local orthonormal vertical basis. Now,

$$\begin{aligned}\widetilde{\mathrm{Ric}}(U,V) &= \widetilde{\mathrm{Ric}}^{\mathbf{h}}(U,V) + \widetilde{\mathrm{Ric}}^{\mathbf{v}}(U,V)\\ &= \sum_i \langle \tilde{R}(X_i,U)V, X_i\rangle_\phi + \sum_j \langle \tilde{R}(\tilde{T}_j,U)V,\tilde{T}_j\rangle_\phi,\end{aligned}$$

where by (2.1.17),

$$\begin{aligned}\langle \tilde{R}(X_i,U)V,X_i\rangle_\phi &= -\langle \tilde{R}(X_i,U)X_i,V\rangle_\phi = -e^{2\phi}\langle \tilde{R}(X_i,U)X_i,V\rangle\\ &= -e^{2\phi}\big(\langle R(X_i,U)X_i,V\rangle + (1-e^{2\phi})\langle A_{X_i}A^*_{X_i}U,V\rangle\\ &\quad + \{h_\phi(X_i,X_i) + \langle \nabla\phi, X_i\rangle^2\}\langle U,V\rangle\\ &\quad - 2\langle \nabla\phi,X_i\rangle\langle S_{X_i}U,V\rangle\big).\end{aligned}$$

Recalling that $\langle S_{X_i}U,V\rangle = \langle \sigma(U,V),X_i\rangle$, we obtain after summing over i,

$$\begin{aligned}\widetilde{\mathrm{Ric}}^{\mathbf{h}}(U,V) = e^{2\phi}\big\{&\mathrm{Ric}^{\mathbf{h}}(U,V) - (1-e^{2\phi})\sum_i \langle A_{X_i}A^*_{X_i}U,V\rangle\\ &- \big(\Delta_M\phi + |\nabla\phi|^2\big)\langle U,V\rangle + 2\langle\nabla\phi,\sigma(U,V)\rangle\big\}.\end{aligned} \tag{2.1.32}$$

Similarly, (2.1.16) implies

$$\begin{aligned}\langle \tilde{R}(\tilde{T}_j,U)V,\tilde{T}_j\rangle_\phi &= \langle \tilde{R}(T_j,U)V,T_j\rangle\\ &= (1-e^{2\phi})\langle R^F(T_j,U)V,T_j\rangle + e^{2\phi}\langle R(T_j,U)V,T_j\rangle\\ &\quad + e^{2\phi}\langle\{(S_{\nabla\phi} - |\nabla\phi|^2 I)(\langle U,V\rangle T_j - \langle T_j,V\rangle U)\},T_j\rangle\\ &\quad + e^{2\phi}\big(\langle\nabla\phi,\sigma(U,V)\rangle\langle T_j,T_j\rangle - \langle\nabla\phi,\sigma(T_j,V)\rangle\langle U,T_j\rangle\big)\\ &= (1-e^{2\phi})\langle R^F(T_j,U)V,T_j\rangle + e^{2\phi}\big\{\langle R(T_j,U)V,T_j\rangle\\ &\quad + \langle U,V\rangle\langle S_{\nabla\phi}T_j,T_j\rangle - \langle T_j,V\rangle\langle S_{\nabla\phi}U,T_j\rangle\\ &\quad - |\nabla\phi|^2\big(\langle U,V\rangle - \langle T_j,V\rangle\langle T_j,U\rangle\big) + \langle\nabla\phi,\sigma(U,V)\rangle\\ &\quad - \langle\nabla\phi,\sigma(T_j,V)\rangle\langle U,T_j\rangle\big\},\end{aligned}$$

so that

$$\begin{aligned}\widetilde{\mathrm{Ric}}^{\mathbf{v}}(U,V) = (1-e^{2\phi})\,\mathrm{Ric}^F(U,V) + e^{2\phi}\big\{&\mathrm{Ric}^{\mathbf{v}}(U,V) + \langle U,V\rangle\big(\langle\nabla\phi,H\rangle\\ &- (k-1)|\nabla\phi|^2\big) + (k-2)\langle\nabla\phi,\sigma(U,V)\rangle\big\}.\end{aligned} \tag{2.1.33}$$

Adding (2.1.32) and (2.1.33) after replacing U, V by T_1, T_2 finally yields

$$\begin{aligned}\widetilde{\mathrm{Ric}}(T_1,T_2) &= (1-e^{2\phi})\big\{\mathrm{Ric}^F(T_1,T_2) - e^{2\phi}\sum_i\langle A_{X_i}A^*_{X_i}T_1,T_2\rangle\big\}\\ &\quad + e^{2\phi}\big\{\mathrm{Ric}(T_1,T_2) + k\langle\nabla\phi,\sigma(T_1,T_2)\rangle\\ &\quad + \langle T_1,T_2\rangle\big(\langle\nabla\phi,H\rangle - \Delta_M\phi - k|\nabla\phi|^2\big)\big\}.\end{aligned} \tag{2.1.34}$$

Example 2.1.1. If $\pi : M \to B$ denotes a Riemannian submersion with totally geodesic fibers, the *canonical variation* of the metric on M is the one obtained by taking the function ϕ to be constant; i.e., $\phi(p) \equiv t$ for all $p \in M$, $t \in (-\infty, \infty)$. The results from Subsection 2.1.3 then imply that the unnormalized sectional curvatures $\tilde{K}$ in the new metric are given by

$$\begin{aligned}\tilde{K}(X,Y) &= (1 - e^{2t})K_B(X,Y) + e^{2t}K(X,Y),\\ \tilde{K}(X,T) &= e^{2t}|A_X^* T|^2,\\ \tilde{K}(T_1,T_2) &= e^{2t}K(T_1,T_2).\end{aligned}$$

In particular, if we let $t \to -\infty$ (which amounts to shrinking the fibers down to a point, a phenomenon known as *collapse*), the sectional curvatures stay bounded, and the metric on M "converges" to that on B.

2.2 Warped products

The simplest application of the construction in Section 2.1 is when the original manifold M is a Riemannian product $B^n \times F^k$, and the Riemannian submersion is the projection onto B. When in addition, the function ϕ depends on B only, then the resulting warped space $(M, \langle , \rangle_\phi)$ is called a *warped product*, and is denoted by $B \times_{e^{2\phi}} F$. By (2.1.9), the A-tensor is identically zero. By (2.1.4), the second fundamental form of the fibers is given by

$$\sigma(T_1,T_2) = -e^{2\phi}\langle T_1,T_2\rangle \nabla\phi = -\langle T_1,T_2\rangle_\phi \nabla\phi.$$

In particular, the fibers are totally umbilic, and the mean curvature field $H = -k\nabla\phi$ is basic. These properties actually characterize warped products:

Proposition 2.2.1. *Let $\pi : M^{n+k} \to B^n$ be a Riemannian submersion. Then π is locally a warped product iff*

1. *the A-tensor identically vanishes;*
2. *the fibers are totally umbilic submanifolds of M, and*
3. *π is isoparametric; i.e., the mean curvature form κ is basic (equivalently, the vector field metrically dual to κ is basic).*

Proof. We have already remarked that the three conditions are necessary, so it remains to establish they are also sufficient. For basic X, Y,

$$d\kappa(X,Y) = -2\,\mathrm{div}\,A_X Y = 0$$

by Proposition 1.4.1, and

$$d\kappa(X,T) = X\kappa(T) - T\kappa(X) - \kappa([X,T]) = 0$$

since κ is basic. Thus, κ is closed, and locally equals $kd\psi$ for some function ψ that is constant on the fibers. Equivalently, $H = k\nabla\psi$. The fibers being totally umbilic means that there is a horizontal vector field N such that $\sigma(T_1, T_2) = \langle T_1, T_2\rangle N$. But for an orthonormal basis T_i, $H = \sum \sigma(T_i, T_i)$, so that $N = (1/k)H = \nabla\psi$. Multiply the metric on the fibers by $e^{2\psi}$. (2.1.4) implies that the second fundamental form of the fibers in the new metric is given by

$$\tilde{\sigma}(T_1, T_2) = e^{2\psi}\{\sigma(T_1, T_2) - \langle T_1, T_2\rangle\nabla\psi\} = 0.$$

Since both the second fundamental tensor and the A-tensor vanish, $(M, \langle,\rangle_\psi)$ is locally a metric product $B\times F$. This means that the original Riemannian manifold is locally a warped product $B \times_{e^{-2\psi}} F$, as claimed. □

In order to construct examples of warped products, we will need the following:

Lemma 2.2.1. *Let M be a Riemannian manifold, $\phi : M \to \mathbb{R}$. Then ϕ is a Riemannian submersion iff $|\nabla\phi| \equiv 1$.*

Proof. Suppose ϕ is a Riemannian submersion, and consider a normal horizontal geodesic of M. Since $\phi \circ c$ is a geodesic of $\mathbb{R}$, $(\phi \circ c)(t) = \pm t + (\phi \circ c)(0)$. But $\nabla\phi$ is horizontal, so that

$$\pm 1 = (\phi \circ c)' = \langle \nabla\phi \circ c, \dot{c}\rangle = \pm|\nabla\phi \circ c|.$$

Conversely, suppose the function $\phi : M \to \mathbb{R}$ has gradient of unit length. Then $\phi_*\nabla\phi(1_\mathbb{R}) = \nabla\phi(1_\mathbb{R} \circ \phi) = \langle\nabla\phi, \nabla\phi\rangle$ =1. Thus, $\phi_*\nabla\phi = D \circ \phi$, where D is the standard coordinate vector field on $\mathbb{R}$, and ϕ_* is a linear isometry on $(\ker \phi_*)^\perp$; i.e., ϕ is a Riemannian submersion. □

Let M be a Riemannian manifold, $p \in M$, and $r_0 > 0$ small enough that $\exp_p$ maps $U_{r_0} := \{v \in M_p \mid |v| < r_0\}$ diffeomorphically onto the metric ball $B_{r_0}(p)$ of radius r_0 centered at p. The function $\phi : B_{r_0}(p) \setminus \{p\} \to (0, r_0)$, where $\phi(q) = d(p, q)$, has maximal rank everywhere, since $\nabla\phi(q)$ is the tangent vector at q of the minimal normal geodesic from p to q. In fact, $|\nabla\phi| \equiv 1$, and by Lemma 2.2.1, ϕ is a Riemannian submersion. Set $X := \nabla\phi$, so that by the proof of the above lemma, X is basic and ϕ-related to the standard coordinate vector field D on $(0, r_0)$. Fix $r \in (0, r_0)$, and consider the sphere $F := \phi^{-1}(r)$ of radius r centered at p. Since X is a unit normal field along F, the second fundamental form of F is given by

$$S_X v = -\nabla_v X = -\nabla_v\nabla\phi, \qquad q \in F, \quad v \in F_q.$$

Notice that we omitted the vertical superscript, since $\nabla_v X$ is already vertical. Let $\gamma : (-\epsilon, \epsilon) \to M_p$ be a smooth curve in the tangent space of p with $\exp_{p*}\dot{\gamma}(0) = v$, $|\gamma| \equiv r$, and consider the variation $V : [0, r] \times (-\epsilon, \epsilon) \to M$ by geodesics, where $V(t, s) = \exp_p\left((t/r)\gamma(s)\right)$. If $c(t) := V(t, 0)$, then by assumption, $V_*D_1 = X \circ V$, and $t \mapsto Y(t) := V_*D_2(t, 0)$ is the Jacobi field along c with $Y(0) = 0$, $Y(r) = v$. Thus,

$$\nabla_v\nabla\phi = \nabla_{D_2(r,0)}V_*D_1 = \nabla_{D_1(r,0)}V_*D_2 = Y'(r),$$

and the second fundamental form of F at a point q is given by $\langle S_X v, w\rangle = -\langle Y', Z\rangle(r)$, where Y and Z are the Jacobi fields along c with $Y(0) = Z(0) = 0$, $Y(r) = v$, $Z(r) = w$.

Suppose now that M is a space of constant curvature λ, and denote by c_λ and s_λ the solutions of the differential equation $x''(t) + \lambda x(t) = 0$ with $c_\lambda(0) = s'_\lambda(0) = 1$, $c'_\lambda(0) = s_\lambda(0) = 0$. The Jacobi fields above are then given by

$$Y = \frac{s_\lambda}{s_\lambda(r)} V, \quad Z = \frac{s_\lambda}{s_\lambda(r)} W,$$

where V, W are the parallel fields along c with $V(r) = v$, $W(r) = w$. Thus,

$$\langle S_X v, w\rangle = -\langle Y', Z\rangle(r) = -\left\langle \frac{c_\lambda}{s_\lambda}(r)v, \frac{s_\lambda}{s_\lambda}(r)w \right\rangle = -\frac{c_\lambda}{s_\lambda}(r)\langle v, w\rangle.$$

It follows that $\kappa(X) = -k(c_\lambda/s_\lambda)\circ\phi$, where $k = \dim M - 1$; equivalently, since $\dot c = X\circ c$,

$$\begin{aligned} \kappa(\dot c) &= -k\frac{c_\lambda}{s_\lambda}\circ\phi\circ c = -k(\ln\circ s_\lambda\circ\phi\circ c)' \qquad \text{(because } \phi\circ c = 1_{(0,r_0)}\text{)} \\ &= k\,d\psi(\dot c), \end{aligned}$$

where $\psi = -\ln\circ s_\lambda\circ\phi$. By the proof of Proposition 2.2.1,

$$B_{r_0}\setminus\{p\} = (0, r_0)\times_{e^{-2\psi}} F_r = (0, r_0)\times_{s_\lambda^2} F_r$$

isometrically, where F_r denotes the sphere $F = \phi^{-1}(r)$ with the induced metric multiplied by $s_\lambda^{-2}(r)$. By the Gauss equations, the sectional curvature of F is $\lambda + (c_\lambda/s_\lambda)^2(r) = (\lambda s_\lambda^2 + c_\lambda^2)/(s_\lambda^2)(r) = 1/(s_\lambda^2)(r)$. Thus, the curvature of F_r is identically 1; i.e., F_r is isometric to the round sphere $S^{n-1}(1)$ of radius 1. Notice that r_0 may be taken to be ∞ if $\lambda \le 0$, and $\pi/\sqrt{\lambda}$ if $\lambda > 0$.

Summarizing, we have proved:

Theorem 2.2.1. *Let Q_λ^n denote the simply connected space form of constant curvature λ, and p a point in Q_λ^n. Then*

1. $Q_\lambda^n\setminus\{p\} = (0,\infty)\times_{s_\lambda^2} S^{n-1}(1)$ *if* $\lambda\le 0$, *and*
2. $Q_\lambda^n\setminus\{p,-p\} = (0,\pi/\sqrt{\lambda})\times_{s_\lambda^2} S^{n-1}(1)$ *if* $\lambda > 0$, *where*

$$s_\lambda(t) = \begin{cases} t & \textit{if } \lambda = 0, \\ \frac{1}{\sqrt{\lambda}}\sin(\sqrt{\lambda}t) & \textit{if } \lambda > 0, \\ \frac{1}{\sqrt{\lambda}}\sinh(\sqrt{\lambda}t) & \textit{if } \lambda < 0. \end{cases}$$

When $\lambda < 0$, all of Q_λ^n can actually be realized as a warped product: Indeed, we may, after rescaling, assume that $\lambda = -1$. Choose some $p \in M^n := Q_{-1}^n$, some $(n-1)$-dimensional subspace N_p of M_p, and set $N := \exp_p(N_p)$. N is a totally geodesic submanifold of M isometric to Q^{n-1}_{-1}. If X is a unit-length section of the normal bundle of N in M, then $f : \mathbb{R} \times N \to M$, $f(t,p) = \exp tX(p)$, is a diffeomorphism because X is parallel and N is totally geodesic. Notice that $f_{*(t_0,p)}(D,0) = \dot{c}(t_0)$, where $c(t) = \exp tX(p)$, so that $|f_{*(t_0,p)}(D,0)| = 1$. Thus, the *signed distance*

$$\pi := \pi_1 \circ f^{-1} : M \to \mathbb{R}$$

from N (where $\pi_1 : \mathbb{R} \times N \to \mathbb{R}$ denotes projection) has gradient of unit length, and is a Riemannian submersion by Lemma 2.2.1. As in the proof of Theorem 2.2.1, the S-tensor of the submersion at $q = \exp t_0 X(p)$ is given by

$$S_{\nabla\pi} v = -\nabla_v \nabla\pi = -Y'(t_0),$$

where Y is the holonomy Jacobi field along c with $Y(t_0) = v$. Now,

$$Y'(0) = -S_{\dot{c}(0)} Y(0) - A^*_{\dot{c}(0)} Y(0) = 0,$$

since N is totally geodesic and the horizontal distribution, being one-dimensional, is integrable. Thus, $Y(t) = (\cosh t/\cosh t_0)V$, where V is the parallel field along c with $V(t_0) = v$, and $S_{\nabla\pi} v = -(\tanh t_0)v$. The mean curvature form κ is therefore given by

$$\begin{aligned}\kappa(\nabla\pi) &= -(n-1)\tanh\circ\pi = -(n-1)(\ln\cosh)'\circ\pi \\ &= -(n-1)(\ln\cosh)'\circ\pi\, d\pi(\nabla\pi) = -(n-1)d(\ln\cosh\circ\pi)(\nabla\pi).\end{aligned}$$

By the proof of Proposition 2.2.1,

$$Q^n_{-1} = \mathbb{R} \times_{\cosh^2} Q^{n-1}_{-1}.$$

There are other realizations of hyperbolic space as warped products: For example, $Q^n_{-1} = \mathbb{R} \times_{e^{2t}} \mathbb{R}^{n-1}$, which comes from fibering Q by horospheres. Similarly, if $\mathbb{R}^+$ denotes $(0,\infty)$ with the complete metric $\phi^2 g_0$, where g_0 is the standard metric and $\phi(t) = 1/t$, then the upper half-space model of Q^n_{-1} is the warped product $\mathbb{R}^+ \times_{\phi^2} \mathbb{R}^{n-1}$.

Let X, Y, Z be basic, T, T_i vertical, $i = 1, 2$. Following the convention adopted in Section 2.1, a basic field on $M = B \times_{e^{2\phi}} F$ will be identified with its π-related vector field on the base B. The curvature identities from Section 2.1 applied to the Riemannian product $B \times F$ immediately yield the following formulas for a warped product $B \times_{e^{2\phi}} F$:

Proposition 2.2.2. *Let R_B, R_F denote the curvature tensors of the Riemannian manifolds B, F, respectively. The curvature R of the warped product $B \times_{e^{2\phi}} F$ is* *then given by*

$$\begin{aligned} R(X,Y)Z &= R_B(X,Y)Z; && (2.2.1)\\ R(T_1,T_2)T_3 &= R_F(T_1,T_2)T_3 - e^{2\phi}|\nabla\phi|^2\big(\langle T_2,T_3\rangle T_1 - \langle T_1,T_3\rangle T_2\big); && (2.2.2)\\ R(X,Y)T &= R(T_1,T_2)X = 0; && (2.2.3)\\ R(X,T)Y &= \big(h_\phi(X,Y) + \langle\nabla\phi,X\rangle\langle\nabla\phi,Y\rangle\big)T; && (2.2.4)\\ R(T_1,X)T_2 &= \langle T_1,T_2\rangle\big(\langle\nabla\phi,X\rangle\nabla\phi + \nabla_X\nabla\phi\big) && (2.2.5) \end{aligned}$$

where the inner products are those in the original product metric.

Similarly, let K, K_B, and K_F denote the non-normalized sectional curvatures of $B\times_{e^{2\phi}}F$, B, and F respectively. Thus, for example, $K(E,F) = \langle R(E,F)F,E\rangle$ in the warped product metric. Proposition 2.2.2 immediately yields:

Corollary 2.2.1.

$$\begin{aligned} K(X,Y) &= K_B(X,Y); && (2.2.6)\\ K(T_1,T_2) &= e^{-2\phi}\{K_F(T_1,T_2) - |\nabla\phi|^2\big(|T_1|^2|T_2|^2 - \langle T_1,T_2\rangle^2\big)\}; && (2.2.7)\\ K(X,T) &= -|T|^2\big(\langle\nabla\phi,X\rangle^2 + h_\phi(X,X)\big). && (2.2.8) \end{aligned}$$

Corollary 2.2.2. *The Ricci curvature of $B\times_{e^{2\phi}}F^k$ satisfies*

$$\begin{aligned} \mathrm{Ric}(X,Y) &= \mathrm{Ric}_B(X,Y) - k\big(h_\phi(X,Y) + \langle\nabla\phi,X\rangle\langle\nabla\phi,Y\rangle\big); && (2.2.9)\\ \mathrm{Ric}(X,T) &= 0; && (2.2.10)\\ \mathrm{Ric}(T_1,T_2) &= \mathrm{Ric}_F(T_1,T_2) - \langle T_1,T_2\rangle e^{2\phi}\big(\Delta_M\phi - k|\nabla\phi|^2\big). && (2.2.11) \end{aligned}$$

Notice that the inner products on the right side of the identities in the above corollaries correspond to the product metric. The following is an immediate consequence of (1.4.17) in Chapter 1:

Proposition 2.2.3. *A curve $c = (c_h, c_v)$ in $B\times_{e^{2\phi}}F$ is a geodesic iff*

$$\nabla_D\dot c_h = e^{2\phi}|\dot c_v|^2\nabla\phi, \qquad \nabla_D\dot c_v = -(\phi\circ c_h)'\dot c_v.$$

We next look at a class of metric foliations that is more general than that of warped products. We have seen that many Riemannian submersions are fiber bundles for which the horizontal distribution actually defines a connection. Recall that a connection on a fiber bundle is *flat* if it is integrable. This motivates the following:

Definition 2.2.1. A metric foliation is said to be *flat* if its horizontal distribution is integrable; equivalently, if $A\equiv 0$.

Examples 2.2.1. (i) A warped product is a flat Riemannian submersion.

(ii) A codimension one metric foliation is necessarily flat. In particular, any Riemannian manifold M *locally* admits flat metric foliations: As pointed out earlier in the section, if $p \in M$ and ϵ is the injectivity radius at p, then

$$\begin{aligned} B_\epsilon(p) \setminus \{p\} &\to (0, \epsilon), \\ q &\mapsto d(p, q) \end{aligned}$$

is a flat Riemannian submersion. A similar statement holds if p is replaced by a compact hypersurface of M.

(iii) Suppose M admits a totally geodesic foliation of codimension one. Its orthogonal complement then defines a one-dimensional flat metric foliation by Definition 1.2.2.

(iv) Let Q^{n+k} be a space of constant curvature, M^k a submanifold of Q with flat normal bundle; i.e., the connection on the normal bundle ν of M, induced by that of Q, is flat. Then *for any $p \in M$, there exists a neighborhood U of p such that $M \cap U$ is a leaf of a flat metric foliation of U.*

To see this, consider the foliation obtained by exponentiating parallel sections of ν, restricted to a neighborhood U small enough to avoid singularities. To establish that this foliation is metric, it suffices, by Definition 1.2.2, to show that each leaf is everywhere orthogonal to the totally geodesic foliation of $M \cap U$ with leaves $\exp(\nu_q)$, $q \in M \cap U$, where ν_q denotes the fiber of ν over q. So consider a leaf $N := \exp X(M)$, where X is a parallel section of ν, and a curve γ in N. Then $\gamma = \exp_c(X \circ c)$ for some curve c in M. Now, the tangent space of $\exp(\nu_{\gamma(0)})$ can be realized as the parallel translate of the tangent space of $\exp(\nu_{c(0)})$ along the geodesic $t \mapsto \exp(t(X \circ c)(0))$. Thus, the claim will follow once we show that if E is a parallel field along $t \mapsto \exp(t(X \circ c)(0))$ with $E(0) \perp M_{c(0)}$, then $E(1) \perp \dot\gamma(0)$. So consider the variation by geodesics

$$(t, s) \mapsto V(t, s) = \exp_{c(s)} t(X \circ c)(s).$$

If J is the Jacobi field $J(t) = V_* D_2(t, 0)$ along the geodesic $t \mapsto V(t, 0)$, then $J(0) = \dot c(0)$, and $J(1) = \dot\gamma(0)$. By hypothesis, $J(0) \perp E(0)$, and

$$J'(0) = \nabla_{D_1(0,0)} V_* D_2 = \nabla_{D_2(0,0)} V_* D_1 = \nabla_{D(0)} (X \circ c)$$

is orthogonal to $E(0)$ since X is a parallel section of ν. But in a space of constant curvature κ, if J is a Jacobi field along a geodesic, E a parallel field along the same geodesic, and both $J(0)$, $J'(0)$ are orthogonal to $E(0)$, then $\langle J, E\rangle \equiv 0$; this is because $\langle J, E\rangle$ satisfies the second-order ODE

$$\langle J, E\rangle'' = \langle J'', E\rangle = -\langle R(J, \dot c)\dot c, E\rangle = -\kappa\langle J, E\rangle,$$

with initial conditions $\langle J, E\rangle = \langle J, E\rangle' = 0$. But then $\langle \dot\gamma(0), E(1)\rangle = \langle J(1), E(1)\rangle = 0$, as claimed.

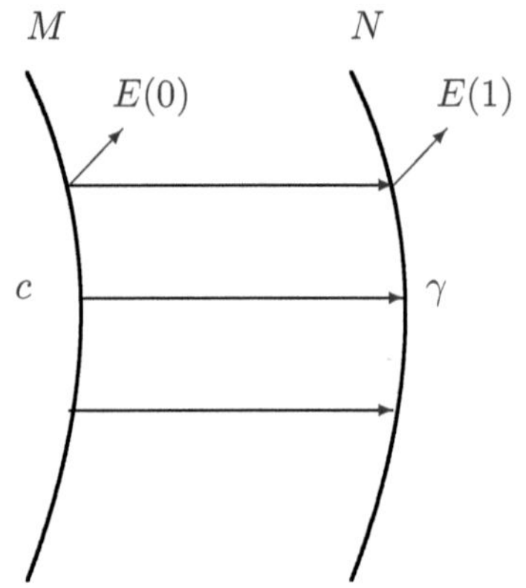

Conversely, of course, if $\mathcal{F}$ is any (local) flat metric foliation in a space form, and M is a leaf of $\mathcal{F}$, then M has flat normal bundle, and any leaf equals $\exp(X(M))$ for some parallel section X of the normal bundle of M: This is in fact already true in nonconstant curvature, since for an arbitrary metric foliation, leaves have the form $\exp(X(M))$ for a *basic* section of the normal bundle. When the foliation is flat, basic sections coincide with the parallel ones by (1.4.4).

The following result is completely general, and is in fact just a reformulation of (1.5.2):

Lemma 2.2.2. *Let $\mathcal{F}$ be a metric foliation, c a horizontal geodesic in M, and denote by S, R^{v}, and A the tensor fields $S_{\dot c}$, $R^{\mathsf{v}}(\cdot, \dot c)\dot c$, and $A_{\dot c}$ respectively along c. One then has the Riccati-type equation:*

$$S'^{\mathsf{v}} = S^2 - AA^* + R^{\mathsf{v}}. \tag{2.2.12}$$

Recall that for unit horizontal X, the vertical Ricci curvature in direction X is $\mathrm{Ric}^{\mathsf{v}}(X,X) = \sum_i K(X,T_i)$, where T_i is an orthonormal basis of the vertical space.

Theorem 2.2.2. *Let $\mathcal{F}$ be a flat metric foliation on a complete Riemannian manifold M. If the vertical Ricci curvature is nonnegative in horizontal directions, then $\mathcal{F}$ splits.*

Proof. Let $c : \mathbb{R} \to M$ be a horizontal geodesic. Since $A \equiv 0$, (2.2.12) becomes $S'^{\mathsf{v}} = S^2 + R^{\mathsf{v}}$. Again by flatness, the vertical distribution $\mathcal{V}$ is invariant under parallel translation along c; similarly, $R(\mathcal{V} \circ c) \subset \mathcal{V} \circ c$. We may therefore identify $\mathcal{V} \circ c$ via parallel translation with the vertical space E at $c(0)$. If $L(E)$ denotes the space of self-adjoint endomorphisms of E, then the Riccati equation becomes

$$S' = S^2 + R, \tag{2.2.13}$$

where $S, R : \mathbb{R} \to L(E)$. The proof of Theorem 1.7.1, with S replaced by $-S$, now goes through essentially word for word to show that both S and R must vanish identically. Since $A \equiv 0$ by assumption, the claim follows. □

Implicit in the proof of Theorem 2.2.2 is the following:

Corollary 2.2.3. *Let M be a complete Riemannian manifold with nonnegative sectional curvature. If the sectional curvatures are positive at some point, then M admits no flat metric foliations.*

Remarks 2.2.1. (i) Theorem 2.2.2 is of course no longer true if the completeness assumption is dropped: We have seen, for example, that Euclidean space with a point deleted can be written as a warped product. Nor is it true if the hypothesis $\mathrm{Ric}^{\mathrm{v}} \geq 0$ is replaced by $\mathrm{Ric} \geq 0$: There exist complete warped product metrics of nonnegative Ricci curvature on $\mathbb{R}^n$, see [5].

(ii) A slightly more involved argument shows that for a flat metric foliation of a complete manifold with sectional curvature bounded below by $-\lambda^2$, the leaves must have principal curvatures no larger than $|\lambda|$ in absolute value, cf. [136]. Observe that when $\lambda = 0$, this implies totally geodesic leaves and therefore splitting of the foliation, see also (iii) below.

(iii) If one replaces the Ricci curvature hypothesis in Theorem 2.2.2 by the stronger condition that M have nonnegative sectional curvature, then the result follows from Examples and remarks 1.8.1 (v). Indeed, if $\mathcal{F}$ is flat, then the dual foliation is the one that is tangent to the horizontal distribution – and has therefore complete leaves, so that any pair of vertical and horizontal vectors generate a totally geodesic flat. In particular, $S \equiv 0$.

(iv) A codimension one metric foliation is automatically flat. By Theorem 2.2.2, such a foliation cannot exist if the Ricci curvature is positive. There are further restrictions, though: if the universal cover $\tilde{M}$ of M is compact, then M does not admit such a foliation. Otherwise, it can be lifted to $\tilde{M}$, and its orthogonal complement is an orientable one-dimensional foliation, so that there exists a global basic unit vector field X. If α denotes the dual one-form,

$$\alpha(E) = \langle X, E\rangle, \qquad E \in \mathfrak{X}M,$$

then for vertical U, V, $d\alpha(U,V) = U\langle X,V\rangle - V\langle X,U\rangle - \langle X,[U,V]\rangle = 0$. Similarly, $d\alpha(U,X) = 0$ because $[X,U]$ is vertical. Thus, α is closed, hence exact, and $\alpha = df$ for some $f : M \longrightarrow \mathbb{R}$. This would imply that α and hence also X vanish at those points where f attains extremal values, which is impossible.

2.3 Homogeneous submersions

The richest class of Riemannian submersions is the one generated by isometric group actions on a Riemannian manifold M. More precisely, let G be a compact Lie group acting by isometries on M via $\mu : G \times M \to M$. For $p \in M$ and $g \in G$, denote by $\imath_p : G \to M$ the map given by $\imath_p(g) = \mu(g, p)$, and by $\jmath_g : M \to M$ the isometry $\jmath_g(p) = \mu(g, p)$ induced by g. When there is no risk of confusion, we also use g instead of $\jmath_g$. The *orbit* $\{g(p) \mid g \in G\}$ of $p \in M$ will be denoted by $G(p)$, and the *isotropy subgroup* G_p at p is the subgroup of G consisting of all $g \in G$ such that $g(p) = p$. Notice that an isotropy subgroup is necessarily closed in G. We assume the action is by *principal orbits*; i.e., for any two orbits O_1 and O_2, there exists a diffeomorphism $f : O_1 \longrightarrow O_2$ that is G-equivariant in the sense that $f \circ \jmath_g = \jmath_g \circ f$ for any $g \in G$.

Lemma 2.3.1. *Let $p \in M$, and set $H := G_p$. Then the map $f : G/H \to M$ given by $f(gH) = g(p)$ is an imbedding onto the orbit $G(p)$ of p.*

Proof. Denote by $\pi : G \longrightarrow G/H$ the projection. f is clearly well defined and bijective onto the orbit of p. Since G/H is compact and M is Hausdorff, f is a topological imbedding, and it remains to establish that it has maximal rank everywhere. But f is equivariant, so it suffices to do so at eH; equivalently, we claim that $x \in G_e$ belongs to H_e whenever $\pi_{*e}x \in \ker f_{*eH}$. To see this, consider the vector field $X \in \mathfrak{g}$ with $X(e) = x$, and the curve $t \mapsto c(t) := (f \circ \pi)(\exp tx)$. Equivariance of f means that $f = \jmath_g \circ f \circ \mathbb{L}_g^{-1}$ for any $g \in G$, where $\mathbb{L}_g : G/H \longrightarrow G/H$ is given by $\mathbb{L}_g(aH) = gaH$. But if L_g denotes left translation in G, then $\mathbb{L}_g \circ \pi = \pi \circ L_g$, so that

$$\begin{aligned}\dot{c}(t) &= (f \circ \pi)_*(X(\exp tx)) = \jmath_{\exp tx*} \circ f_* \circ \mathbb{L}_{\exp -tx*}\pi_* X(\exp tx)\\ &= \jmath_{\exp tx*} \circ f_* \circ \pi_* \circ L_{exp-tx*}X(\exp tx) = \jmath_{\exp tx*} \circ f_*\pi_* x = 0.\end{aligned}$$

$\exp tx$ therefore belongs to H for all t, and $x \in H_e$.

□

Remark 2.3.1. When G is not assumed to be compact, Lemma 2.3.1 no longer holds in general, even when the action is free. If a is an irrational number, then the $\mathbb{R}$-action on the torus $S^1 \times S^1$ given by $\mu(t, (z_1, z_2)) = (z_1 e^{it}, z_2 e^{iat})$ is free and isometric, but all orbits are dense.

Lemma 2.3.2. *The space $G \backslash M$ of orbits inherits a natural differentiable structure from M such that the projection $\pi : M \to G\backslash M$ becomes a submersion. It also inherits a natural metric for which π becomes Riemannian.*

Proof. Consider a point $p \in M$ with isotropy group $H = G_p$. Let ν_p denote the normal bundle of $G(p)$ in M, ν_p^ϵ the corresponding disk bundle of radius $\epsilon > 0$, and U the fiber of the latter over p. Choose ϵ small enough so that $\exp : E(\nu_p^\epsilon) \longrightarrow B_\epsilon(G(p))$ is a diffeomorphism from the total space of the disk bundle onto the tubular neighborhood of radius ϵ of the orbit, and consider the map $\phi : (G/H) \times U \longrightarrow B_\epsilon(G(p))$ given by $\phi(gH, x) = g(\exp x)$. To see that this map is well defined, we must show that H acts trivially on U; i.e., that $h(\exp x) = \exp x$ for all $h \in H$ and $x \in U$. By hypothesis, there exists an equivariant diffeomorphism $\psi : G/G_{\exp x} \longrightarrow G/H$. Let $aH = \psi(G_{\exp x})$. By equivariance, $\psi(gG_{\exp x}) = gaH$

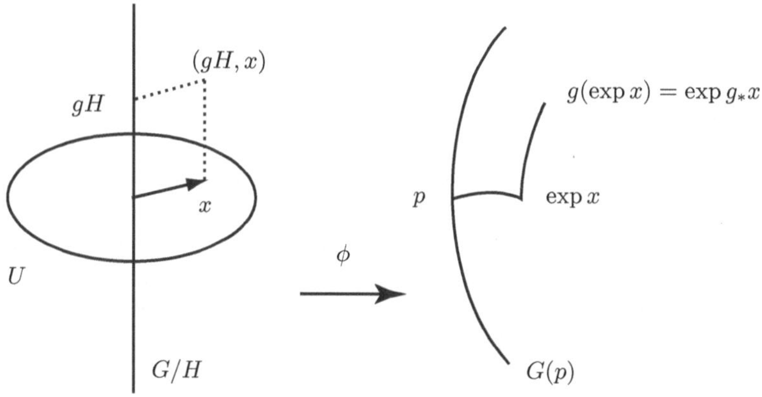

for $g \in G$. This implies that for any $g \in G_{\exp x}$, $gaH = aH$, and $a^{-1}ga \in H$. In other words, the isotropy subgroup of $\exp x$ is conjugate to a subgroup of H. Arguing in a similar fashion with ψ^{-1}, we see that it is conjugate to H itself. But if $g \in G_{\exp x}$, then $\exp x = g(\exp x) = \exp g_* x$ (because g is an isometry), so that $g_* x = x$, and in particular $g(p) = p$; i.e., $g \in H$. Thus, $G_{\exp x} = H$, and H acts trivially on U as claimed.

ϕ is a G-equivariant diffeomorphism. Its inverse, followed by projection onto the second factor yields a homeomorphism between a neighborhood of $G(p)$ (in $G\backslash M$ endowed with the quotient topology) and U. It is easy to check that the transition functions between two such homeomorphisms are smooth, so that $G\backslash M$ inherits a differentiable structure for which the projection $\pi : M \to G\backslash M$ is differentiable. As noted in Chapter 1, since the action is by isometries, there exists a unique Riemannian metric on the quotient for which π becomes Riemannian. □

Notice that the proof actually establishes that π is a fibration. We will shortly see that it is in fact a fiber bundle.

Set $B := G\backslash M$, and for $b \in B$ denote by $\mathrm{Hol}(b)$ the holonomy group of the submersion at b, consisting of the holonomy diffeomorphisms of $\pi^{-1}(b)$ obtained by horizontally lifting piece-wise smooth loops at b.

Lemma 2.3.3. $\mathrm{Hol}(b)$ *is a Lie group.*

Proof. Consider a curve $c : [0,1] \to B$, and the diffeomorphism

$$h_c : \pi^{-1}(c(0)) \to \pi^{-1}(c(1))$$

between the fibers over the endpoints that assigns to p the endpoint of the horizontal lift of c starting at p. If $\bar{c}$ is a horizontal lift of c, then $\jmath_g \circ \bar{c}$ is a horizontal curve for any $g \in G$ because $\jmath_g$ is an isometry, and by definition of π, it projects down to c. This means that h_c is a G-equivariant diffeomorphism. Thus, $\mathrm{Hol}(b)$ is a subgroup of the group $\mathrm{Diff}_G(\pi^{-1}(b))$ of G-equivariant diffeomorphisms of $\pi^{-1}(b)$. We claim that the latter group is a Lie group. To see this, identify $\pi^{-1}(b)$ with G/H as in Lemma 2.3.1. It will suffice to establish that $\mathrm{Diff}_G(G/H)$ is isomorphic to $N(H)/H$, where $N(H)$ is the normalizer of H in G. Now, if $f \in \mathrm{Diff}_G(G/H)$, then $f(gH) = gaH$ for some $a \in G$ with $a^{-1}Ha \subset H$ by the argument used in Lemma 2.3.2. We claim that $a^{-1}Ha = H$; i.e., $a \in N(H)$: Indeed, if $A = \{a^n \mid n = 0, 1, 2, \dots\}$, then by Lemma 2.3.4 below, the closure $\bar{A}$ of A contains a^{-1}. Now, the map $F : G \times G \to G$ sending (b, c) to $b^{-1}cb$ is continuous, and by hypothesis, $F(A \times H) \subset H$. Since H is closed, $F(\bar{A} \times H) \subset H$. Thus, $aHa^{-1} \subset H$, so that $H \subset a^{-1}Ha$ as claimed. Summarizing, any $f \in \mathrm{Diff}_G(G/H)$ has the form $f(gH) = gaH = (gH)a$ for some a in the normalizer of H. It now easily follows that $\mathrm{Diff}_G(G/H)$ is isomorphic to $N(H)/H$ acting by right multiplication on G/H.

Having established that $\mathrm{Diff}_G(\pi^{-1}(b))$ is a Lie group, consider the subgroup $\mathrm{Hol}_0(b)$ of $\mathrm{Hol}(b)$ obtained by considering only loops that are null-homotopic. This is a path-connected subgroup of $\mathrm{Diff}_G(\pi^{-1}(b))$, and therefore also a Lie group, see, e.g., [142]. There is a natural epimorphism of $\pi_1(B, b)$ onto $\mathrm{Hol}(b)/\mathrm{Hol}_0(b)$ that assigns to the homotopy class of a loop c the equivalence class of h_c. Since the fundamental group of B is countable, so is $\mathrm{Hol}(b)/\mathrm{Hol}_0(b)$. Thus, $\mathrm{Hol}(b)$ is a Lie group, as claimed. □

Lemma 2.3.4. *For any $a \in G$, the closure of the set $A = \{a^n \mid n = 0, 1, 2, \dots\}$ is a subgroup of G.*

Proof. Notice first of all that the closure of a subgroup is again a subgroup by continuity of $(a, b) \mapsto ab^{-1}$. It suffices therefore to show that $a^{-1} \in \bar{A}$, or equivalently, that any neighborhood of a^{-1} intersects A. Consider the subgroup $\langle a \rangle$ generated by a. If e is an isolated point of $\overline{\langle a \rangle}$, then $\overline{\langle a \rangle}$ is discrete, and being compact, must be finite, so that $a^n = e$ for some $n \in \mathbb{N}$. If $n = 1$, then $a^{-1} = e \in A$, and otherwise, $a^{-1} = a^{n-1} \in A$. So assume that e is not isolated. If U is a neighborhood of e, then so is $V = U \cap U^{-1}$, where $U^{-1} := \{g^{-1} \mid g \in U\}$. It must therefore contain a^n for some positive n, so that $a^{n-1} \in L_{a^{-1}}(V) \cap A$. In other words, if U is any neighborhood of e, then $L_{a^{-1}}(U)$ intersects A. But then any neighborhood W of a^{-1} intersects A, because $L_a(W)$ is a neighborhood of e, so that $W \cap A = L_{a^{-1}}(L_a(W)) \cap A \neq \emptyset$. □

Theorem 2.3.1. *Let G be a compact Lie group acting by isometries on M with principal orbits, so that $\pi : M \to B := G \backslash M$ is a Riemannian submersion. Fix any $b_0 \in B$, let $F = \pi^{-1}(b_0)$, and* $\mathrm{Hol}(b_0)$ *be the holonomy group of the submersion at b_0. Then M is the total space of a fiber bundle with fiber F and structure group* $\mathrm{Hol}(b_0)$.

Proof. Consider a locally finite cover of B by open sets U_α each of which is the diffeomorphic image via exp of some metric ball in the tangent space of $b_\alpha \in U_\alpha$. For each α, choose a geodesic c_α from b_0 to b_α. Given $b \in B_\alpha$, there exists a unique minimal normal geodesic c_α^b from b_α to b. This yields a trivialisation

$\{\pi^{-1}(U_\alpha),(\pi,\phi_\alpha)\}$ of the fibration π, where $(\pi,\phi_\alpha)^{-1}: U_\alpha \times F \to \pi^{-1}(U_\alpha)$ is given by $(\pi,\phi_\alpha)^{-1}(b,p) = h_{c_\alpha^b} \circ h_{c_\alpha}(p)$.

Here, as above, h_c denotes the holonomy transformation along the curve c. The transition function $\phi_{\alpha\beta}: U_\alpha \cap U_\beta \to \mathrm{Hol}(b_0)$ between two charts U_α and U_β maps $b \in U_\alpha \cap U_\beta$ to the element of $\mathrm{Hol}(b_0)$ that consists of horizontal lifts of the piece-wise smooth curve $c_\beta * c_\beta^b * -c_\alpha^b * -c_\alpha$. Since $\phi_{\alpha\beta}$ depends differentiably on b, the claim follows. □

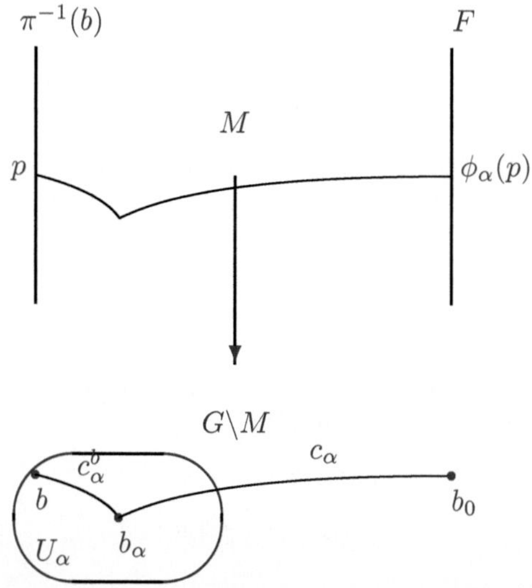

In the case when M is nonnegatively curved, and positively curved at one point, it follows from Wilking's dual foliation result that the fiber is a Lie group:

Proposition 2.3.1 ([73]). *Let M be as in Theorem 2.3.1. If the curvature of M is nonnegative, and positive at some point, then the isotropy group at any point is a normal subgroup of G. Thus, every fiber is diffeomorphic to a Lie group. If furthermore, G is simple, then the action is free or transitive.*

Proof. The argument used in Lemma 2.3.2 implies that any two points that are joined by a horizontal geodesic share the same holonomy group. By Theorem 1.8.1, the leaf of the dual foliation passing through a point where the curvature is positive must have the same dimension as M (since its normal space along a horizontal geodesic is spanned by parallel Jacobi fields). Furthermore, it can be shown that when the action is by isometries, dual leaves are intrinsically complete. Thus, there is only one dual leaf, and the isotropy group is the same for all points of M. But if H is the isotropy group at p, then the isotropy group at $g(p)$ is gHg^{-1}. This implies that H is normal in G. The last two statements follow immediately. □

In the case when M has strictly positive curvature in the above proposition, Wilking has shown that the action is always free [140].

Recall from the proof of Lemma 2.3.3 that $\mathrm{Hol}(b_0)$ acts on F *from the right.* To make this more explicit, denote by $h_c \in \mathrm{Hol}(b_0)$ the holonomy diffeomorphism of F induced by the loop c at b_0. Given loops c_1 and c_2 at b_0, $c_1 * c_2$ is the loop obtained by first following c_1, and then c_2. The group multiplication $*$ in $\mathrm{Hol}(b_0)$ is then given by $h_{c_1} * h_{c_2} := h_{c_1 * c_2}$. Since $h_{c_1 * c_2} = h_{c_2} \circ h_{c_1}$, the right action of $\mathrm{Hol}(b_0)$ on F is just $(p, h) \mapsto h(p)$.

Set $Q := \mathrm{Hol}(b_0)$, and consider the principal Q-bundle $\pi_P : P \to B$ associated to $\pi : M \to B$. Since Q acts on F from the right, π_P is a left principal bundle, and there exists a G-equivariant diffeomorphism

$$\Phi : F \times_Q P \longrightarrow M$$

from the total space $F \times_Q P$ of the associated F-bundle with M (recall that $F \times_Q P$ is the orbit space of $F \times P$ under the left Q-action given by $h(p, f) = (ph^{-1}, hf)$, for $h \in Q$, $p \in F$, and $f \in P$).

The total space P of the principal bundle may be described as follows: for $b \in B$, let Q_b denote the collection of all holonomy transformations $h_c : F \to \pi^{-1}(b)$, where c is a piece-wise smooth curve from b_0 to b. Define P to be the union of all Q_b as b ranges over B, and $\pi_P : P \to B$ the projection that assigns to an element of Q_b the point b. There is a natural *left* action

$$\begin{aligned} \nu : Q \times P &\to P, \\ (h, f) &\mapsto f \circ h \end{aligned}$$

of Q on P by composition, since

$$\nu(h_1 * h_2, f) = f \circ (h_1 * h_2) = f \circ h_2 \circ h_1 = \nu(h_1, f \circ h_2) = \nu(h_1, \nu(h_2, f)).$$

The bundle atlas $\{\pi^{-1}(U_\alpha), (\pi, \phi_\alpha)\}$ constructed for $\pi : M \to B$ in the proof of Theorem 2.3.1 induces a corresponding atlas $\{\pi_P^{-1}(U_\alpha), (\pi_P, \psi_\alpha)\}$ for π_P, where $\psi_\alpha : \pi_P^{-1}(U_\alpha) \to Q$ maps $f \in Q_b$ to $h_{-c_\alpha^b * -c_\alpha} \circ f$ in the notation of the proof of Theorem 2.3.1. This is a principal bundle atlas, since ψ_α is Q-equivariant: if $h \in Q$, then

$$\psi_\alpha(hf) = h_{-c_\alpha^b * -c_\alpha} \circ f \circ h = \psi_\alpha(f) \circ h = h * \psi_\alpha(f).$$

It is easy to see that the transition functions coincide with those of the F-bundle π, so that π_P is indeed the corresponding principal bundle. If $[p, f] \in F \times_Q P$ denotes the equivalence class of $(p, f) \in F \times P$, then the G-equivariant diffeomorphism $\Phi : F \times_Q P \longrightarrow M$ is given by $\Phi[p, f] = f(p)$.

Recall that a *connection* on a (left) principal Q-bundle $\pi_P : P \to B$ is a distribution $\mathcal{H}$ on P such that $TP = \ker \pi_{P*} \oplus \mathcal{H}$, and $\jmath_{g*}\mathcal{H} = \mathcal{H} \circ \jmath_g$ for all $g \in Q$, where $\jmath_g : P \to P$ maps p to $\nu(g, p) = gp$. *Parallel translation* in the bundle along a curve $c : [0, 1] \to B$ is the diffeomorphism $\mathcal{P}_c : \pi_P^{-1}(c(0)) \to \pi_P^{-1}(c(1))$ between the fibers over the endpoints of c defined by $\mathcal{P}_c(p) = \gamma(1)$, where γ is the horizontal lift of c (i.e., $\dot\gamma \in \mathcal{H} \circ \gamma, \pi_P \circ \gamma = c$) with $\gamma(0) = p$. The *holonomy group*

$\mathrm{Hol}(b)$ of the connection at a point $b \in B$ is the group of all diffeomorphisms of the fiber $\pi_P^{-1}(b)$ over b consisting of parallel translation along piece-wise smooth loops beginning and ending at b. Since B is connected, any two holonomy groups are isomorphic. In our situation, there exists a canonical connection on $P \to B$ induced by the horizontal distribution of the Riemannian submersion $\pi : M \to B$: Consider a point $p = h_c \in P$ and a curve $c_0 : [\alpha, \beta] \to B$ in B with $c_0(\alpha) = \pi_P(p)$. The curve $c_0^p : [\alpha, \beta] \to P$, where

$$c_0^p(t) := h_{c_0|_{[\alpha,t]}} \circ h_c = h_{c * c_0|_{[\alpha,t]}} \tag{2.3.1}$$

is then a lift of c_0 starting at p. Furthermore, given any $m \in F$, the curve $\bar{c}_0$, where

$$\bar{c}_0 = \Phi[m, c_0^p],$$

is by construction the horizontal lift to M (for the Riemannian submersion $\pi : M \to B$) of c_0 with initial condition $\bar{c}_0(\alpha) = \Phi[m, p] = h_c(m)$.

Denote by $\rho : F \times P \to F \times_Q P$ the projection that maps (m, p) to its equivalence class $[m, p]$. By the above, the restriction

$$\Phi_{*[m,p]}|_{\rho_*(\{0_m\} \times P_p)} : \rho_*(\{0_m\} \times P_p) \to M_{\Phi[m,p]}$$

is onto $\mathcal{H}_{\Phi[m,p]}$, so that there exists a unique distribution $\tilde{\mathcal{H}}$ on P such that $(\Phi \circ \rho)_*(\{0_m\} \times \tilde{\mathcal{H}}_p) = \mathcal{H}_{\Phi[m,p]}$. Now, if $\Phi[m, c_0^p]$ is a horizontal curve in M, then so is $\Phi[mg, c_0^p]$ for any $g \in Q$. But the latter is just $\Phi[m, \jmath_g \circ c_0^p]$. Thus, $\jmath_{g*}\tilde{\mathcal{H}} = \tilde{\mathcal{H}} \circ \jmath_g$, and $\tilde{\mathcal{H}}$ is indeed a connection on P. It is easy to describe the holonomy group $\widetilde{\mathrm{Hol}}(b_0)$ of this connection at the point $b_0 \in B$ that was used in the construction of P: By definition, the fiber $\pi_P^{-1}(b_0)$ over b_0 is the group of all holonomy transformations $h_c : F = \pi^{-1}(b_0) \to F$ along loops c based at b_0; i.e., $\pi_P^{-1}(b_0)$ is just $Q = \mathrm{Hol}(b_0)$. By (2.3.1), $\widetilde{\mathrm{Hol}}(b_0) = \mathrm{Hol}(b_0) = Q$ acting on itself by *right* multiplication

$$\begin{aligned} \mathrm{Hol}(b_0) \times \widetilde{\mathrm{Hol}}(b_0) &\to \mathrm{Hol}(b_0), \\ (h_c, h_{c_0}) &\mapsto h_{c * c_0} = h_c * h_{c_0}. \end{aligned}$$

Summarizing the above discussion, we have:

Theorem 2.3.2. *Consider the Riemannian submersion $\pi : M \to B$ from Theorem* 2.3.1, *and the corresponding principal Q-bundle $\pi_P : P \to B$, where $Q = \mathrm{Hol}(b_0)$. Let $\rho : F \times P \to F \times_Q P \simeq M$ denote the projection. If $\mathcal{H}$ denotes the horizontal distribution of the submersion π, then the distribution $\tilde{\mathcal{H}}$ on P determined by $\rho_*(\{0\}_m \times \tilde{\mathcal{H}}_p) = \mathcal{H}_{\rho(m,p)}$, $m \in F$, $p \in P$, is a connection on the principal bundle. The holonomy group of this connection at b_0 is the holonomy group Q of the Riemannian submersion π acting on itself by right multiplication.*

This justifies to a certain extent the terminology introduced in the last section: a homogeneous metric foliation is flat in the sense of Definition 2.2.1 iff the corresponding connection from Theorem 2.3.2 is flat.

We next consider the important special case when the action of G on M is free. If we identify $F = \pi^{-1}(b_0)$ with G via the map from Lemma 2.3.1, then $M = G \times_Q P \to B$ is a left principal G-bundle. Notice that $Q = \mathrm{Hol}(b_0)$ is now identified with a subgroup of G acting on G by right multiplication: Suppose the identification $G \cong F$ is given by $g \mapsto g(p)$, for some $p \in M$. If $h \in Q$ maps p to $\tilde{g}(p)$ for some $\tilde{g} \in G$, then $h = R_{\tilde{g}}$, because

$$gh = h(g(p)) = g(h(p)) = g\tilde{g}.$$

(For the second equality, we have used the fact that holonomy transformations are G-equivariant, as observed in the proof of Lemma 2.3.3.) Thus, the bundle $P \to B$ is the reduction of $M \to B$ to a principal Q-bundle. The decomposition $TM = \ker \pi_* \oplus \mathcal{H}$ of the tangent bundle of M induced by the connection induces one at the vector level, and we write $u = u^{\mathbf{v}} + u^{\mathbf{h}} \in \ker \pi_* \oplus \mathcal{H}$ as usual. If $\imath_p$ denotes the imbedding of G onto the fiber containing $p \in M$ from Lemma 2.3.1, then the *connection form* of $\mathcal{H}$ is the $\mathfrak{g}$-valued one-form ω on M given by

$$\omega(u) = (\imath_{p*e})^{-1} u^{\mathbf{v}}, \qquad u \in M_p, \quad p \in M,$$

and the *curvature form* of $\mathcal{H}$ is the $\mathfrak{g}$-valued 2-form Ω defined by

$$\Omega(x, y) = -\omega[X, Y]^{\mathbf{v}}(p), \qquad x, y \in M_p, \quad p \in M, \tag{2.3.2}$$

where X, Y are local horizontal fields on M with $X(p) = x^{\mathbf{h}}$, $Y(p) = y^{\mathbf{h}}$. It follows that after identification of the vertical space with the Lie algebra $\mathfrak{g}$ of the group, the A-tensor of the Riemannian submersion is equal to $-(1/2)\Omega$.

A well-known result of Ambrose and Singer states that the Lie algebra of the holonomy group of a connection on a principal G-bundle is the Lie subalgebra of $\mathfrak{g}$ generated by the image of Ω. We summarize all these facts in the following:

Proposition 2.3.2. *Let $\pi : M \to B = G\backslash M$ denote the Riemannian submersion generated by a free isometric action of a compact Lie group G on M. Then the holonomy group of the submersion is a Lie group, and its Lie algebra is the subalgebra of $\mathfrak{g}$ generated by*

$$\{\imath_{p*e}^{-1} A_x y \mid x, y \in \mathcal{H}_p, \, p \in M\},$$

where $\imath_p : G \to M$ is the imbedding $g \mapsto g(p)$ onto the fiber containing p.

Definition 2.3.1. The *fundamental vector field* $\tilde{U}$ on M induced by $U \in \mathfrak{g}$ is given by

$$\tilde{U}(p) := \imath_{p*} U(e). \tag{2.3.3}$$

Recall that a vector field on M is said to be *Killing* if its flow consists of isometries of M.

Proposition 2.3.3. *The collection of fundamental vector fields on M is a Lie algebra of Killing fields that is isomorphic to the algebra of* right-*invariant fields on G.*

Proof. Let $U \in \mathfrak{g}$, and $\tilde{U}$ be the corresponding fundamental vector field. To see that $\tilde{U}$ is Killing, it suffices to show that its flow ϕ_t is given by $\jmath_{\exp tU}$. So consider the curve c, where $c(t) = \jmath_{\exp tU}(q)$, and $q \in M$. Given $t_0 \in \mathbb{R}$,

$$c(t) = \jmath_{\exp tU} q = \imath_q(\exp tU) = \imath_q(\exp(t-t_0)U \exp t_0 U) = \imath_{c(t_0)}(\exp(t-t_0)U).$$

Thus, $\dot{c}(t_0) = \imath_{c(t_0)*}U = \tilde{U} \circ c(t_0)$, as claimed, and $\tilde{U}$ is Killing. For the second statement, let $\bar{\mathfrak{g}}$ denote the Lie algebra of right-invariant fields on G. Given $U \in \mathfrak{g}$, $\bar{U} \in \bar{\mathfrak{g}}$ with $U(e) = \bar{U}(e)$, and $p \in M$, the identity $\imath_{gp} = \imath_p \circ R_g$ implies that

$$\tilde{U}(gp) = \imath_{gp*}U(e) = \imath_{p*}R_{g*}U(e) = \imath_{p*}\bar{U}(g),$$

so that $\tilde{U} \circ \imath_p = \imath_{p*}\bar{U}$, and $\imath_{p*}$ is a Lie algebra isomorphism. □

The vector fields $A_X Y$ are not, in general, Killing fields. This is because they are associated to left-invariant fields of G, whereas the fundamental Killing fields are associated to right-invariant ones:

Definition 2.3.2. A vector field on a fiber F is said to be *left-invariant* if it is $\jmath_g$-related to itself for any $g \in G$.

The collection of left-invariant fields on a fiber form a Lie algebra isomorphic to $\mathfrak{g}$. In fact, given u in the tangent space of F at some p, it is straightforward to see that the left-invariant field $\hat{U}$ on F that equals u at p is given by $\hat{U} = \imath_{p*} \circ U \circ \imath_p^{-1}$, where U is the element of $\mathfrak{g}$ with $U(e) = \imath_{p*}^{-1}u$.

Proposition 2.3.4. *Given a left-invariant $\hat{U}$ and basic X, Y along a fiber F, the vector fields $A_X Y$ and $S_X \hat{U}$ are left-invariant.*

Proof. A horizontal field X is basic iff it is $\jmath_g$-related to itself for any $g \in G$. Thus, for $g \in G$, basic X, and left-invariant $\hat{U}$,

$$\jmath_{g*}\nabla_{\hat{U}} X = \nabla_{\hat{U}} \jmath_{g*} X = \nabla_{\hat{U}}(X \circ \jmath_g) = \nabla_{\jmath_{g*}\hat{U}} X = (\nabla_{\hat{U}} X) \circ \jmath_g,$$

since the map $\jmath_g$ is an isometry. For the same reason, this map preserves the vertizontal splitting, so that

$$\jmath_{g*} S_X \hat{U} = (S_X \hat{U}) \circ \jmath_g, \qquad \jmath_{g*} A_X^* \hat{U} = (A_X^* \hat{U}) \circ \jmath_g.$$

This implies that $S_X \hat{U}$ is left-invariant, and $A_X^* \hat{U}$ is basic. In particular, for basic Y, $\langle A_X Y, \hat{U} \rangle = \langle A_X^* \hat{U}, Y \rangle$ is constant, so that $A_X Y$ is left-invariant. □

Example 2.3.1. Consider the free $\mathbb{R}^k$-action on $\mathbb{R}^{k+n} = \mathbb{R}^k \times \mathbb{R}^n$ given by

$$\begin{aligned} \mathbb{R}^k \times (\mathbb{R}^k \times \mathbb{R}^n) &\to \mathbb{R}^k \times \mathbb{R}^n, \\ (v, (u, x)) &\mapsto (u + v, \phi(v)x) \end{aligned}$$

where $\phi : \mathbb{R}^k \to SO(n)$ is a Lie group homomorphism. The orbits of the action are generalized helices winding around the central plane $\mathbb{R}^k \times \{0\}$, which happens to be the orbit of the origin. Here, G is the abelian group $\mathbb{R}^k$, so the left-invariant fields coincide with the right-invariant ones, and are then necessarily parallel along a fiber F, since they form an abelian Lie algebra of Killing fields that contains a point-wise orthonormal basis U_i: in fact, $[U_i, U_j] = 0$, so that

$$\begin{aligned}
2\langle \nabla_{U_i} U_j, U_k \rangle &= \langle \nabla_{U_i} U_j, U_k \rangle + \langle \nabla_{U_j} U_i, U_k \rangle \\
&= -\langle \nabla_{U_k} U_j, U_i \rangle - \langle \nabla_{U_k} U_i, U_j \rangle \\
&= -U_k \langle U_i, U_j \rangle = 0,
\end{aligned}$$

and $\langle U_i, U_j \rangle$ is constant along F.

2.4 Left-invariant metrics on Lie groups

The simplest and most common homogeneous submersions are those from Lie groups with left-invariant metrics. Recall that such a metric on a Lie group G is one for which each left translation $L_g : G \to G$, where $L_g(a) = ga$, is an isometry. Such metrics are in bijective correspondence with inner products on the tangent space G_e of G at the identity e by letting the canonical isomorphism

$$\begin{aligned}
\mathfrak{g} &\to G_e, \\
X &\mapsto X(e)
\end{aligned}$$

from the Lie algebra $\mathfrak{g}$ of G be a linear isometry. In the sequel, we will often use this isomorphism to identify both spaces. We begin by computing the connection and curvature of such a metric. Since it is left-invariant, the formula

$$\begin{aligned}
\langle \nabla_X Y, Z \rangle = \frac{1}{2}\{ & X\langle Y, Z \rangle + Y\langle Z, X \rangle - Z\langle X, Y \rangle \\
& + \langle Z, [X, Y] \rangle + \langle Y, [Z, X] \rangle - \langle X, [Y, Z] \rangle \}
\end{aligned}$$

for the Levi-Civita connection, when applied to left-invariant vector fields X, Y, $Z \in \mathfrak{g}$ becomes

$$\langle \nabla_X Y, Z \rangle = \frac{1}{2}\{ \langle Z, [X, Y] \rangle + \langle Y, [Z, X] \rangle - \langle X, [Y, Z] \rangle \},$$

so that

$$\nabla_X Y = \frac{1}{2}\{ \operatorname{ad}_X Y - \operatorname{ad}_X^* Y - \operatorname{ad}_Y^* X \}, \qquad X, Y \in \mathfrak{g}, \tag{2.4.1}$$

with ad^* denoting the adjoint of ad. A long, but fairly straightforward computation using (2.4.1) yields for the curvature tensor

$$\begin{aligned} R(X,Y)Z = -\frac{1}{4}\{ &\operatorname{ad}_{[X,Y]} Z + (\operatorname{ad}_X - \operatorname{ad}_X^*)\operatorname{ad}_Z^* Y - (\operatorname{ad}_Y - \operatorname{ad}_Y^*)\operatorname{ad}_Z^* X \\ &+ (\operatorname{ad}_X^* \operatorname{ad}_Y - (\operatorname{ad}_X^* \operatorname{ad}_Y)^*)Z - (\operatorname{ad}_Y \operatorname{ad}_X^* - (\operatorname{ad}_Y \operatorname{ad}_X^*)^*)Z \\ &- \operatorname{ad}_{[X,Y]}^* Z + \operatorname{ad}_{[Y,Z]}^* X + \operatorname{ad}_{[Z,X]}^* Y - \operatorname{ad}^*_{(\operatorname{ad}_Y^* Z + \operatorname{ad}_Z^* Y)} X \\ &+ \operatorname{ad}^*_{(\operatorname{ad}_X^* Z + \operatorname{ad}_Z^* X)} Y\} + \frac{1}{2}\operatorname{ad}_Z^*[X,Y]. \end{aligned}$$

This in turn implies that the non-normalized sectional curvature $K(X,Y) = \langle R(X,Y)Y, X\rangle$ is given by

$$\begin{aligned} K(X,Y) = \frac{1}{2}\langle[[X,Y],X],Y\rangle - \frac{1}{2}\langle[[X,Y],Y],X\rangle - \langle \operatorname{ad}_Y^* Y, \operatorname{ad}_X^* X\rangle \\ - \frac{3}{4}|[X,Y]|^2 + \frac{1}{4}|\operatorname{ad}_X^* Y + \operatorname{ad}_Y^* X|^2. \end{aligned} \tag{2.4.2}$$

Example 2.4.1. Consider the Lie group $G = \mathbb{R}^{n-1} \times \mathbb{R}^+$ with multiplication

$$(x,t)\cdot(y,s) := (x+ty, ts), \qquad x,y \in \mathbb{R}^{n-1}, \quad t,s>0.$$

Left translation $L_{(x,t)}$ by (x,t), when extended to all of $\mathbb{R}^n$, is the affine transformation $(y,s) \mapsto (x,0) + t(y,s)$, so that for the standard coordinate vector fields D_i on $\mathbb{R}^n$, the derivative of $L_{(x,t)}$ at the identity $e = (0,1)$ satisfies

$$L_{(x,t)*e} D_i = tD_i(x,t).$$

It follows that $X_1, \dots, X_n$, where $X_i = tD_i$, is a basis of the Lie algebra of G, and

$$[X_i, X_j] = 0, \quad [X_n, X_i] = X_i, \qquad i,j < n.$$

Endow G with the left-invariant metric for which X_i is an orthonormal basis. By the above,

$$\operatorname{ad}_{X_i}^* X_j = -\delta_{ij} X_n, \quad , \operatorname{ad}_{X_i}^* X_n = \operatorname{ad}_{X_n}^* X_n = 0, \quad \operatorname{ad}_{X_n}^* X_i = X_i.$$

(2.4.2) then implies

$$K(X_i, X_j) = -\langle \operatorname{ad}_{X_i}^* X_i, \operatorname{ad}_{X_j}^* X_j\rangle = -1,$$

$$\begin{aligned} K(X_i, X_n) &= -\frac{1}{2}\langle[[X_i,X_n],X_n],X_i\rangle - \frac{3}{4}|[X_i,X_n]|^2 + \frac{1}{4}|\operatorname{ad}_{X_n}^* X_i|^2 \\ &= -\frac{1}{2} - \frac{3}{4} + \frac{1}{4} = -1. \end{aligned}$$

Thus, G is a simply connected space of constant curvature -1. Being homogeneous, it is also complete, and is therefore isometric to hyperbolic space. In fact, the identity map is an isometry between G and the upper half-space model of hyperbolic space.

Lemma 2.4.1. *If* $\operatorname{ad}_X : \mathfrak{g} \to \mathfrak{g}$ *is skew-adjoint, then*

$$K(X,Y) = \frac{1}{4}|\operatorname{ad}_Y^* X|^2 \geq 0$$

for all $Y \in \mathfrak{g}$, *and* $K(X,Y) = 0$ *iff* $X \perp \operatorname{ad}_Y(\mathfrak{g})$.

Proof. By skew-symmetry of ad_X,

$$\langle [[X,Y],X],Y\rangle = |\operatorname{ad}_X Y|^2, \qquad \langle [[X,Y],Y],X\rangle = -\langle \operatorname{ad}_X Y, \operatorname{ad}_Y^* X\rangle,$$
$$\langle \operatorname{ad}_Y^* Y, \operatorname{ad}_X^* X\rangle = 0, \qquad |\operatorname{ad}_X^* Y + \operatorname{ad}_Y^* X|^2 = |-\operatorname{ad}_X Y + \operatorname{ad}_Y^* X|^2.$$

The claim follows upon substitution of these expressions in (2.4.2). □

Proposition 2.4.1. *Let H be a connected Lie subgroup of G. The following statements are equivalent:*

1. *Right translation $R_h : G \to G$ by h, $R_h(g) = gh$, is an isometry for any $h \in H$.*
2. *$\mathfrak{h}$ is an algebra of Killing fields of G.*
3. $\operatorname{Ad}_h : \mathfrak{g} \to \mathfrak{g}$ *is a linear isometry for any $h \in H$.*
4. $\operatorname{ad}_X : \mathfrak{g} \to \mathfrak{g}$ *is skew-adjoint for any $X \in \mathfrak{h}$.*

Proof. For any $X \in \mathfrak{g}$, the curve $t \mapsto \exp(tX(e))$ is the integral curve of X passing through e when $t = 0$ (here, $\exp : \mathfrak{g} \to G$ denotes the Lie group exponential map). By left invariance of X, $t \mapsto L_g(\exp tX(e))$ is the integral curve of X passing through g at 0. The identity $R_{\exp tX}(g) = L_g(\exp tX)$ then implies that the flow of X is given by $R_{\exp tX}$; this shows the equivalence of (1) and (2). Since $\operatorname{Ad}_{h^{-1}} = L_{h^{-1}*} \circ R_{h*e}$, and since $L_{h^{-1}*}$ is a linear isometry, the equivalence of (1) and (3) is clear. The equivalence of (3) and (4) need only be established in a neighborhood of e, because H is connected. Choose one such that is the diffeomorphic image via exp of some open neighborhood of 0 in H_e. If $h = \exp x$, then ad_x is skew-adjoint iff $\operatorname{ad}_x^* = -\operatorname{ad}_x$; this occurs iff $e^{\operatorname{ad}_x^*} = e^{-\operatorname{ad}_x}$, or equivalently, $(e^{\operatorname{ad}_x})^* = (e^{\operatorname{ad}_x})^{-1}$. Using the identity

$$\operatorname{Ad} \circ \exp = e^{\operatorname{ad}},$$

we see that this amounts to the condition that $\operatorname{Ad}_h^* = (\operatorname{Ad}_h)^{-1}$; i.e., that Ad_h is a linear isometry. □

It is well known that if G acts effectively on G/H, then the statements in the above proposition are satisfied (for some left-invariant metric on G) iff $\operatorname{Ad}(H)$ has compact closure in $GL(\mathfrak{g})$, cf. [35]. In particular, any compact Lie group admits a *bi-invariant metric*; i.e., a metric for which both left and right translations are isometries. Notice that by Lemma 2.4.1 and Proposition 2.4.1, if for $X \in \mathfrak{g}$, right translation by $\exp tX$ is an isometry for all t, then $K(X,Y) \geq 0$ for all $Y \in \mathfrak{g}$. For example, a Lie group with bi-invariant metric has nonnegative sectional curvature

$$K(X,Y) = \frac{1}{4}|[X,Y]|^2, \qquad X,Y \in \mathfrak{g}. \tag{2.4.3}$$

On the other hand, many of these curvatures will be zero: Wallach has shown that S^3 is the only simply connected Lie group that admits a left-invariant metric with strictly positive curvature [129]. The fact that bi-invariant metrics have nonnegative curvature can also be seen geometrically [43]: let M be a Riemannian manifold, and suppose that for any pair c_1, c_2 of normal geodesics with $c_1(0) = c_2(0)$, the distance between $c_1(2t)$ and $c_2(2t)$ is at most twice that between $c_1(t)$ and $c_2(t)$ for small enough t. Then M has nonnegative curvature by the Rauch comparison theorem, see also [9]. To establish this for a Lie group G with bi-invariant metric,

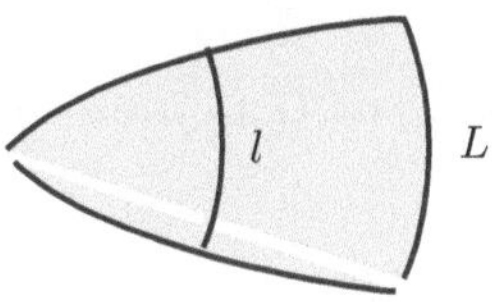

$L \leq 2l \Rightarrow K \geq 0$

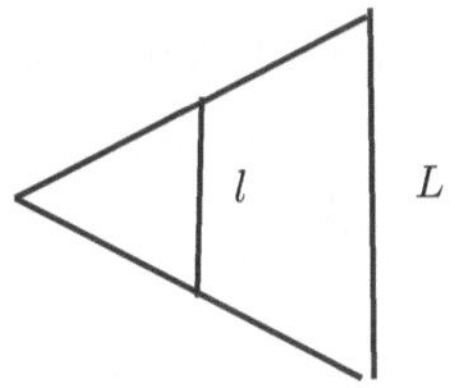

$L = 2l \Rightarrow K = 0$

consider a pair c_1, c_2 of normal geodesics of G with $c_1(0) = c_2(0)$. By homogeneity, we may assume they have e as initial point. Fix any small t, and consider $g = c_1(t)$, $h = c_2(t)$. Then

$$\begin{aligned} d(g^2, h^2) &= d(L_{h^{-1}} \circ R_{g^{-1}}(g^2), L_{h^{-1}} \circ R_{g^{-1}}(h^2)) = d(h^{-1}g, hg^{-1}) \\ &\leq d(h^{-1}g, e) + d(e, hg^{-1}). \end{aligned}$$

Next, apply $L_{h^{-1}} \circ R_g$ to both elements in the last distance term to deduce that $d(g^2, h^2) \leq 2d(h^{-1}g, e)$. Finally, we obtain

$$d(g^2, h^2) \leq 2d(h^{-1}g, e) = 2d(L_h(h^{-1}g), L_h(e)) = 2d(g, h).$$

Since c_1 is a one-parameter subgroup of G, $g^2 = c_1(t)^2 = c_1(2t)$, and similarly, $h^2 = c_2(2t)$. This establishes the claim.

Let us examine in more detail the curvature tensor of a bi-invariant metric. By (2.4.1),

$$\nabla_X = \frac{1}{2} \operatorname{ad}_X,$$

so that

$$\begin{aligned} R(X, Y) &= \nabla_X \nabla_Y - \nabla_Y \nabla_X - \nabla_{[X,Y]} \\ &= \frac{1}{4}(\operatorname{ad}_X \circ \operatorname{ad}_Y - \operatorname{ad}_Y \circ \operatorname{ad}_X) - \frac{1}{2} \operatorname{ad}_{[X,Y]}. \end{aligned}$$

By the Jacobi identity, $\operatorname{ad}_{[X,Y]} = \operatorname{ad}_X \circ \operatorname{ad}_Y - \operatorname{ad}_Y \circ \operatorname{ad}_X$, and

$$R(X, Y) = -\frac{1}{4} \operatorname{ad}_{[X,Y]}. \tag{2.4.4}$$

Given a Riemannian metric on a manifold M, define the *Ricci form* ρ of the metric by

$$\rho(x) := \operatorname{tr} R_x, \qquad x \in M_p, \quad p \in M,$$

where R_x is the self-adjoint endomorphism of M_p given by $R_x(y) = R(y,x)x$. Observe that $\operatorname{Ric}(x,x) = \rho(x)$. We claim that the Ricci form is independent of the particular bi-invariant metric on G: recall that the *Killing form* of $\mathfrak{g}$ is the bilinear form β, where

$$\beta(X,Y) = \operatorname{tr}(\operatorname{ad}_X \circ \operatorname{ad}_Y), \qquad X, Y \in \mathfrak{g}.$$

The claim then follows from:

Lemma 2.4.2. *The Ricci form ρ of a bi-invariant metric on G is given by $\rho(X) = -(1/4)\beta(X,X)$.*

Proof. For $X \in \mathfrak{g}$, (2.4.4) implies that $R_X = -(1/4)(\operatorname{ad}_X)^2$, since

$$R_X(Y) = -\frac{1}{4}\operatorname{ad}_{[Y,X]} X = -\frac{1}{4}\operatorname{ad}_X [X,Y]. \qquad \square$$

It follows from the lemma that the Ricci tensor of a bi-invariant metric satisfies

$$\operatorname{Ric} = -\frac{1}{4}\beta. \tag{2.4.5}$$

We now look at an important special case: recall that a Lie algebra is said to be *simple* if it is nonabelian and has no proper ideals. A Lie algebra is *semisimple* if it is the direct sum of simple ideals. Thus, for example, a semisimple Lie algebra must have trivial center, since the latter is an abelian ideal. A Lie group is said to be simple or semisimple if its Lie algebra has these properties.

Proposition 2.4.2. *If β denotes the Killing form of a compact semisimple Lie group G, then $-\beta$ is a bi-invariant metric on G, called the* canonical metric *of G. The canonical metric is Einstein with Ricci curvature* 1/4.

Proof. The Killing form of any Lie algebra is symmetric and bilinear; we claim it is negative definite when G is semisimple. Consider any bi-invariant metric on G (such a metric exists by compactness of the group), and a corresponding orthonormal basis X_i of $\mathfrak{g}$. Since ad_X is skew-adjoint for this metric,

$$\beta(X,X) = \operatorname{tr}(\operatorname{ad}_X)^2 = \sum_i \langle \operatorname{ad}_X^2 X_i, X_i \rangle = -\sum_i |\operatorname{ad}_X X_i|^2 \leq 0,$$

and can only be zero if $\operatorname{ad}_X = 0$. In that case, X belongs to the center of $\mathfrak{g}$, which is trivial. Thus, $-\beta$ is an inner product on $\mathfrak{g}$. We claim it generates a bi-invariant metric on G; i.e., that ad_X is skew-symmetric with respect to β for any $X \in \mathfrak{g}$, so that

$$\beta([X,Y],Z) = -\beta(Y,[X,Z]), \qquad Y, Z \in \mathfrak{g}.$$

But this is a consequence of the following calculation:

$$\begin{aligned}\operatorname{tr}(\operatorname{ad}_{[X,Y]} \circ \operatorname{ad}_Z) &= \operatorname{tr}(\operatorname{ad}_X \circ \operatorname{ad}_Y \circ \operatorname{ad}_Z - \operatorname{ad}_Y \circ \operatorname{ad}_X \circ \operatorname{ad}_Z)\\ &= -\operatorname{tr}(\operatorname{ad}_Y \circ \operatorname{ad}_X \circ \operatorname{ad}_Z - \operatorname{ad}_X \circ \operatorname{ad}_Y \circ \operatorname{ad}_Z)\\ &= -\operatorname{tr}(\operatorname{ad}_Y \circ \operatorname{ad}_X \circ \operatorname{ad}_Z - \operatorname{ad}_Y \circ \operatorname{ad}_Z \circ \operatorname{ad}_X)\\ &= -\operatorname{tr}(\operatorname{ad}_Y \circ \operatorname{ad}_{[X,Z]}).\end{aligned}$$

Thus $-\beta$ is a bi-invariant metric on G, as claimed. Its Ricci curvature is given by (2.4.5). □

When G is actually simple, the canonical metric is essentially the only bi-invariant metric on G:

Proposition 2.4.3. *If G is a compact simple Lie group, then the canonical metric is the only bi-invariant metric, up to scaling by some constant.*

Proof. We follow Milnor's argument [87]. Any other bi-invariant metric $\langle , \rangle$ on G can be expressed as $\langle X, Y\rangle = -\beta(LX, Y)$ for some self-adjoint $L : \mathfrak{g} \to \mathfrak{g}$. Given $Z \in \mathfrak{g}$, ad_Z is skew-adjoint with respect to both metrics, and

$$\begin{aligned}-\beta(L \operatorname{ad}_Z X, Y) = \langle \operatorname{ad}_Z X, Y\rangle = -\langle X, \operatorname{ad}_Z Y\rangle &= \beta(LX, \operatorname{ad}_Z Y)\\ &= -\beta(\operatorname{ad}_Z LX, Y),\end{aligned}$$

so that ad_Z and L commute. Eigenspaces of L are then invariant under ad_Z for any $Z \in \mathfrak{g}$, which means that they are in fact ideals. Since G is simple, L can have only one eigenvalue λ, and $\langle X, Y\rangle = -\lambda \cdot \beta(X, Y)$. □

We now return to the general case of an arbitrary left-invariant metric on G. If H is a subgroup of G, then the left action

$$\begin{aligned} H \times G &\to G,\\ (h, g) &\mapsto hg\end{aligned}$$

of H on G is by isometries, and the collection of orbits is a metric foliation on G. If, in addition, H is closed, then the orbit space $H\backslash G = \{Hg \mid g \in G\}$ admits a manifold structure for which the projection $\pi : G \to H\backslash G$ is a submersion, and there exists a unique metric on the quotient such that π is a Riemannian submersion. Notice that if H is normal in G, then the orbit space $H\backslash G = G/H$ is a Lie group.

Example 2.4.2. The *Heisenberg algebra* is the $(2n+1)$-dimensional Lie algebra $\mathfrak{h}_n$, where the only nontrivial bracket operations on a given basis $X_1, Y_1, \dots, X_n, Y_n, Z$ are

$$[X_i, Y_i] = -[Y_i, X_i] = Z, \qquad i = 1, \dots, n.$$

Clearly, $\mathfrak{h}_n$ has one-dimensional center $\mathfrak{z}$ spanned by Z. The *Heisenberg group* H_n is the simply connected Lie group that has $\mathfrak{h}_n$ as its Lie algebra. The one-dimensional subgroup $\mathbb{R} = \{\exp(tZ) \mid t \in \mathbb{R}\}$ is normal in H_n, so that $H_n/\mathbb{R}$ is a Lie group, and for any left-invariant metric on H_n, there exists a metric on the quotient such that $\pi : H_n \to H_n/\mathbb{R}$ is a Riemannian submersion. Fix one such metric. There exists an orthonormal basis of $\mathfrak{h}_n$, which we denote by the same letters as before, in which the only nontrivial bracket relations are given by

$$[X_i, Y_i] = -[Y_i, X_i] = 2\alpha_i Z, \qquad i = 1, \dots, n, \tag{2.4.6}$$

for some nonzero $\alpha_i \in \mathbb{R}$. By (2.4.1), $\nabla_X Z = -(1/2)\operatorname{ad}_X^* Z$ for X in the Lie algebra of H_n, which, together with (2.4.6), yields

$$\nabla_Z X_i = \nabla_{X_i} Z = -\alpha_i Y_i, \qquad \nabla_Z Y_i = \nabla_{Y_i} Z = \alpha_i X_i.$$

Similarly,

$$\nabla_{X_i} Y_i = -\nabla_{Y_i} X_i = \alpha_i Z,$$

and all other covariant derivatives vanish. By (2.4.2) and (2.4.3),

$$K(X_i, Y_i) = -3\alpha_i^2, \qquad K(X_i, Z) = K(Y_i, Z) = \alpha_i^2,$$

and the sectional curvatures of planes spanned by other pairs of vectors in this orthonormal basis are zero. It follows that the curvature of the quotient space $H_n/\mathbb{R}$ is

$$K(X_i, Y_i) + \frac{3}{4}|[X_i, Y_i]|^2 = 0.$$

In fact, it is easily checked that the Lie group $H_n/\mathbb{R}$ is just the abelian group $\mathbb{R}^{2n}$.

Returning to the general case, for arbitrary H and a given left-invariant metric on G, the fibration of G by left cosets gH, $g \in G$, will not, in general, be metric. However, if $\operatorname{Ad}(H)$ has compact closure, then there exists a left-invariant metric on G that is right-invariant under H, and the fibration is homogeneous for this metric, since a fiber gH is the set $\{R_h(g) \mid h \in H\}$, and each R_h is an isometry. Furthermore, if G/H is endowed with that metric for which $\pi : G \to G/H$ becomes a Riemannian submersion, then the natural action

$$G \times G/H \to G/H,$$
$$(g, aH) \mapsto gaH$$

of G on $M := G/H$ is by isometries: If we denote by $\mathbb{L}_g : M \to M$ the action of $g \in G$, then $\mathbb{L}_g \circ \pi = \pi \circ L_g$, so that $\mathbb{L}_g$ is an isometry of M. M is then called a (Riemannian) *homogeneous space*. The submersion $H_n \longrightarrow \mathbb{R}^{2n}$ described in Example 2.4.2 above falls in this category.

The fundamental tensors of the submersion $\pi : G \to M$ are entirely determined by their values at a single point (which we may therefore choose to be the identity): Indeed, both horizontal and vertical distributions are invariant under left translation, so that if x, $y \in \mathcal{H}_e$, $u \in \mathcal{V}_e$, and X, Y, U denote the left-invariant vector fields that equal x, y, u respectively at e, then $X(g)$, $Y(g) \in \mathcal{H}_g$, and $U(g) \in \mathcal{V}_g$ for any $g \in G$. By (2.4.1), $A_X Y$ and $S_X U$ are then left-invariant.

Let $\mathfrak{m} := \mathfrak{h}^\perp$ denote the orthogonal complement of the Lie algebra $\mathfrak{h}$ of H in $\mathfrak{g}$, so that any $Z \in \mathfrak{g}$ decomposes as $Z = Z_\mathfrak{h} + Z_\mathfrak{m} \in \mathfrak{h} \oplus \mathfrak{m}$. The elements of $\mathfrak{m}$ then span the horizontal distribution at any point, and given X, $Y \in \mathfrak{m}$, $A_X Y = \frac{1}{2}[X,Y]_\mathfrak{h}$. Together with (2.4.2), this implies that the sectional curvature of the plane P spanned by orthonormal $\pi_* X$ and $\pi_* Y$ in TM is given by

$$K(P) = \frac{1}{2}\langle[[X,Y],X],Y\rangle - \frac{1}{2}\langle[[X,Y],Y],X\rangle - \langle \operatorname{ad}_X^* X, \operatorname{ad}_Y^* Y\rangle \\ + \frac{1}{4}|\operatorname{ad}_X^* Y + \operatorname{ad}_Y^* X|^2 - \frac{3}{4}|[X,Y]_\mathfrak{m}|^2.$$

The second fundamental tensor σ of the fibers is

$$\sigma(U,V) = \frac{1}{2}\{(\operatorname{ad}_U V)_\mathfrak{m} - (\operatorname{ad}_U^* V)_\mathfrak{m} - (\operatorname{ad}_V^* U)_\mathfrak{m}\}, \qquad U, V \in \mathfrak{h}.$$

But $\operatorname{ad}_U V$ belongs to $\mathfrak{h}$ since $\mathfrak{h}$ is an algebra, and so does $\operatorname{ad}_U^* V = -\operatorname{ad}_U V$. Thus, the fibers are totally geodesic, and their intrinsic curvature equals the one in G. Since the restriction of the metric to $\mathfrak{h}$ is bi-invariant, the curvature of the fibers is given by (2.4.3). Furthermore, for orthonormal $X \in \mathfrak{m}$, $U \in \mathfrak{h}$, $K(X,U) = (1/4)|\operatorname{ad}_X^* U|^2$. Summarizing, we have proved:

Theorem 2.4.1. *Let G be a Lie group, H a closed subgroup of G, and consider a left-invariant metric on G that is right-invariant under H. Then*

1. *there exists a unique metric on $M := G/H$ such that the projection $\pi : G \to M$ becomes a Riemannian submersion;*
2. *G acts by isometries on M in this metric via $g(aH) = (ga)H$, so that M is a homogeneous space;*
3. *the fibers of π are totally geodesic;*
4. *for orthonormal X, $Y \in \mathfrak{m} := \mathfrak{h}^\perp$, U, $V \in \mathfrak{h}$,*

$$A_X Y = \frac{1}{2}[X,Y]_\mathfrak{h}, \quad K(U,V) = \frac{1}{4}|\operatorname{ad}_U V|^2, \quad K(X,U) = \frac{1}{4}|\operatorname{ad}_X^* U|^2,$$

and $K(X,Y)$ is given by (2.4.2).

An important special case is that of a bi-invariant metric on G. G/H with the induced metric is then called a *normal homogeneous space*. By the above theorem and (2.4.3), we have:

Corollary 2.4.1. *A normal homogeneous space has nonnegative sectional curvature. Specifically, for orthonormal X, $Y \in \mathfrak{m}$,*

$$K(\pi_* X, \pi_* Y) = \frac{1}{4}|[X,Y]_\mathfrak{m}|^2 + |[X,Y]_\mathfrak{h}|^2.$$

Example 2.4.3 (The Berger spheres). Recall that the unit sphere in $\mathbb{R}^4$ is the Lie group of unit quaternions under quaternion multiplication. Consider the Lie group $S^3 \times \mathbb{R}$ with the bi-invariant metric that is the product of the standard bi-invariant metrics on each factor. If $c : \mathbb{R} \to S^3$ is a unit-speed one-parameter subgroup of S^3, and α is a positive number less than 1, then

$$H := \{(c(\alpha t), \sqrt{1-\alpha^2}t) \in S^3 \times \mathbb{R} \mid t \in \mathbb{R}\}$$

is a subgroup of $S^3 \times \mathbb{R}$, and the manifold $M = (S^3 \times \mathbb{R})/H$ inherits a normal homogeneous metric from the projection $\pi : S^3 \times \mathbb{R} \to M$.

Choose an orthonormal basis X_1, X_2, X_3 of the Lie algebra of S^3 with $\dot{c}(0) = X_1(e)$, and $[X_i, X_{i+1}] = 2X_{i+2} \pmod 3$, see for example [136]. The vertical distribution of $\pi : S^3 \times \mathbb{R} \to M$ is spanned by $T := (\alpha X_1, \sqrt{1-\alpha^2}D)$, and the horizontal distribution by $X := (-\sqrt{1-\alpha^2}X_1, \alpha D)$, $(X_2, 0)$, and $(X_3, 0)$. Since T spans the kernel of π_*, the restriction $\pi : S^3 \times \{0\} \to M$ of π has maximal rank everywhere, and is therefore a local diffeomorphism. It is also injective, so that M is diffeomorphic to S^3.

We compute the curvature of M. Any pair Y, Z of horizontal vectors may be expressed as

$$Y = a_1 X + a_2(X_2, 0) + a_3(X_3, 0), \qquad Z = b_1 X + b_2(X_2, 0) + b_3(X_3, 0).$$

Then a straightforward computation yields

$$[Y, Z] = 2c_1(X_1, 0) - \sqrt{1-\alpha^2}(2c_2(X_2, 0) + 2c_3(X_3, 0)),$$

where $c_1 = a_2 b_3 - a_3 b_2$, $c_2 = a_3 b_1 - a_1 b_3$, $c_3 = a_1 b_2 - a_2 b_1$. It follows that

$$|[Y, Z]|^2 = 4(c_1^2 + (1-\alpha^2)(c_2^2 + c_3^2)).$$

Furthermore, $|Y|^2|Z|^2 - \langle Y, Z\rangle^2 = c_1^2 + c_2^2 + c_3^2$, and

$$|[Y, Z]_{\mathfrak{h}}|^2 = \langle [Y, Z], T\rangle^2 = 4c_1^2\alpha^2.$$

By Corollary 2.4.1, the curvature of the plane P spanned by $\pi_* Y$ and $\pi_* Z$ is

$$\begin{aligned} K_P &= \frac{|[Y,Z]|^2 + 3|[Y,Z]_{\mathfrak{h}}|^2}{4(|Y|^2|Z|^2 - \langle Y, Z\rangle^2)} = \frac{c_1^2 + (1-\alpha^2)(c_2^2 + c_3^2) + 3c_1^2\alpha^2}{c_1^2 + c_2^2 + c_3^2} \\ &= 1 + \alpha^2 \frac{3c_1^2 - (c_2^2 + c_3^2)}{c_1^2 + c_2^2 + c_3^2}, \end{aligned}$$

so that the curvature K_M of M satisfies

$$0 < 1 - \alpha^2 \le K_M \le 1 + 3\alpha^2.$$

It is a well-known fact that an even-dimensional positively curved manifold with curvature bounded above by κ has injectivity radius $\ge \pi/\sqrt{\kappa}$. This example by Berger shows that the result no longer holds in odd dimensions: Indeed, we claim that M has a closed geodesic of length $2\pi\sqrt{1-\alpha^2}$. Since $\pi\sqrt{1-\alpha^2} < \pi/(\sqrt{1+3\alpha^2})$ provided $\alpha^2 > 2/3$, the injectivity radius of M must be smaller than $\pi/(\sqrt{1+3\alpha^2})$. To establish the claim, observe that by (2.4.1), c is a geodesic of S^3 with period 2π. The curve γ, with $\gamma(t) = \bigl(c(-\sqrt{1-\alpha^2}t), \alpha t\bigr)$, is then a geodesic in $S^3 \times \mathbb{R}$ that is horizontal, since its tangent vector at 0 is $X(e)$. Furthermore,

$$\gamma(2\pi\sqrt{1-\alpha^2}) = \bigl(c(-2\pi(1-\alpha^2)), 2\pi\alpha\sqrt{1-\alpha^2}\bigr) = \bigl(c(2\pi\alpha^2), 2\pi\alpha\sqrt{1-\alpha^2}\bigr),$$

by periodicity of c. Thus, $\gamma(2\pi\sqrt{1-\alpha^2}) \in H$, and $\pi \circ \gamma$ is a closed geodesic in M of length $2\pi\sqrt{1-\alpha^2}$. This establishes the claim.

Example 2.4.4. The Berger spheres in the previous example may be viewed as a special case of the vertical warping discussed in Section 2.1, and is closely related to the following construction due to Cheeger [34]: Let G be a compact group of isometries of a Riemannian manifold M. Suppose M has nonnegative sectional curvature, and endow G with a bi-invariant metric. G acts freely by isometries on the Riemannian product $G \times M$ by left multiplication

$$\bar{g}(g,p) = (\bar{g}g, \bar{g}(p)), \qquad \bar{g}, g \in G, p \in M,$$

and the quotient space is diffeomorphic to M via

$$\begin{aligned} (G \times M)/G &\longrightarrow M, \\ [(g,m)] &\longmapsto g^{-1}(m). \end{aligned}$$

The new metric on M therefore also has nonnegative curvature. It can be described as the original metric on M shrunk in the direction of the G-orbits. To see this, denote, as in Section 2.3, by $\tilde{U}$ the fundamental Killing field on M induced by $U \in \mathfrak{g}$: $\tilde{U}(p) = \imath_{p*}U(e)$ for $p \in M$, where $\imath_p : G \to M$ maps $g \in G$ to $g(p) \in M$. After identification of the tangent space of $G \times M$ at (e,p) with $G_e \times M_p$, the vertical space of the submersion $\pi : G \times M \to M$ is spanned by the collection

$$(U(e), \tilde{U}(p)), \qquad U \in \mathfrak{g}. \tag{2.4.7}$$

It follows that if $x \in M_p$ is orthogonal to the orbit $G(p)$, then $(0,x)$ is horizontal and $\pi_*(0,x) = x$. Thus, the length of vectors orthogonal to the G-orbits is unchanged, as claimed. Notice that for $U \in \mathfrak{g}$,

$$\pi_*(U(e), 0(p)) = -\tilde{U}(p), \tag{2.4.8}$$

as can be seen by differentiating the identity

$$\pi(\exp(tU), p) = (\exp tU)^{-1}(p) = \imath_p \circ \exp t(-U).$$

(2.4.7) and (2.4.8) now imply that

$$\pi_*(U(e), \tilde{V}(p)) = -\tilde{U}(p) + \tilde{V}(p), \qquad U, V \in \mathfrak{g}. \tag{2.4.9}$$

To see what happens to vectors tangent to orbits, consider $U \in \mathfrak{g}$. By (2.4.7) and (2.4.9), the horizontal lift of $\tilde{U}(p)$ to $G \times M$ at (e, p) is

$$(0, \tilde{U})(e, p) - \langle (0, \tilde{U}), (U, \tilde{U}) \rangle \frac{(U, \tilde{U})}{|(U, \tilde{U})|^2}(e, p).$$

Thus, one easily computes that the length of $\tilde{U}(p)$ squared in the new metric equals

$$\frac{|\tilde{U}|^2}{1 + (|\tilde{U}|^2/|U|^2)}(p) < |\tilde{U}|^2(p).$$

Notice that if we scale the inner product on $\mathfrak{g}$ by $1/t$ and denote by g_t the resulting metric on the quotient $(G \times M)/G$, $t \in (0, 1]$, then g_1 is the metric just described, g_t converges to the original metric when $t \to 0$, and g_t has nonnegative curvature for all t.

The metric on $\mathbb{R}^2$ constructed in Examples and Remarks 1.5.1(iv) is of the type just described, with $G = S^1$. The Killing field $\tilde{U}$ in this case is the polar coordinate field $\partial/\partial\theta$, and by the above, the metric in polar coordinates is given by $dr^2 + (r^2/(1 + r^2))d\theta^2$. The surface is asymptotic to a cylinder of radius 1.

2.5 The Aloff-Wallach examples

Apart from the rank one symmetric spaces (namely spheres, complex and quaternionic projective spaces, and the Cayley plane), examples of compact manifolds with positive sectional curvature are very scarce. In this section, we describe an infinite family of seven-dimensional manifolds of positive curvature following [2]. They are all Riemannian homogeneous spaces, although the metric is not normal homogeneous in the sense of Section 2.4.

Denote by $U(n)$ the unitary group of $n \times n$ complex matrices A such that $A\bar{A}^t = I_n$, and by $SU(n)$ the subgroup consisting of those matrices with determinant 1. Any $A \in U(n)$ may be written as $A = \exp X$ for some X in the Lie algebra of $U(n)$. Then

$$I = \exp X \exp \bar{X}^t,$$

so that $\exp \bar{X}^t = (\exp X)^{-1} = \exp(-X)$, and $\bar{X}^t = -X$ for small X. Conversely, if $\bar{X}^t = -X$, then for $A := \exp X$,

$$A\bar{A}^t = (\exp X)(\exp \bar{X}^t) = \exp(X + \bar{X}^t) = I,$$

and $A \in U(n)$. Similarly, the identity

$$e^{\operatorname{tr} X} = \det(\exp X)$$

implies that $\exp X \in SU(n)$ iff $\operatorname{tr} X = 0$. Thus, $U(n)$ and $SU(n)$ are Lie groups of dimension n^2 and $n^2 - 1$ respectively, with Lie algebras

$$\mathfrak{u}(n) = \{X \in M_{n,n}(\mathbb{C}) \mid X + \bar{X}^t = 0\}, \qquad \mathfrak{su}(n) = \{X \in \mathfrak{u}(n) \mid \operatorname{tr} X = 0\}.$$

Consider the Lie group $G = SU(3)$. The inner product

$$\langle X, Y \rangle := -\operatorname{Re}\operatorname{tr}(XY)$$

on $\mathfrak{su}(3)$ is Ad-invariant, and therefore induces a bi-invariant metric on G. Denote by

$$K = \left\{ \begin{bmatrix} A & 0 \\ 0 & \det A^{-1} \end{bmatrix} \mid A \in U(2) \right\}$$

the standard imbedding of $U(2)$ in G, with Lie algebra

$$\mathfrak{k} = \left\{ \begin{bmatrix} X & 0 \\ 0 & -\operatorname{tr} X \end{bmatrix} \mid X \in \mathfrak{u}(2) \right\}.$$

By the remark following Theorem 2.4.1, there is a normal homogeneous metric on G/K such that the projection $G \to G/K$ is a Riemannian submersion with totally geodesic fibers. In fact, G/K is $\mathbb{C}P^2$ with its canonical metric, and (G, K) is a symmetric pair: the map that assigns to $gK \in G/K$ the complex line containing $g\mathbf{e}_3$ (with $\mathbf{e}_3 = (0,0,1) \in \mathbb{C}^3$) is a diffeomorphism $G/K \longrightarrow \mathbb{C}P^2$. Our next objective is to deform the metric on G: Let G_ϕ denote G with the metric from Section 2.1 warped in the vertical direction by the number $e^{2\phi(t)}$, where $\phi(t) = \ln\sqrt{1+t}$, $t > -1$. The results from Sections 2.1 and 2.4 imply that the curvature tensor $\tilde{R}$ of G_ϕ is given by

$$\begin{aligned}
\langle \tilde{R}(X,Y)Y, X\rangle_\phi &= \frac{1}{4}|[X,Y]|^2 - \frac{3t}{4}|[X,Y]_{\mathfrak{k}}|^2, \\
\langle \tilde{R}(X,T)T, X\rangle_\phi &= \frac{1+t}{4}|[X,T]|^2, \\
\langle \tilde{R}(T_1,T_2)T_2, T_1\rangle_\phi &= \frac{1+t}{4}|[T_1,T_2]|^2
\end{aligned} \tag{2.5.1}$$

for X, $Y \in \mathfrak{k}^\perp$, T, $T_i \in \mathfrak{k}$, with $Z_{\mathfrak{k}}$ denoting the orthogonal projection of $Z \in \mathfrak{g}$ onto $\mathfrak{k}$.

Since

$$\mathfrak{k}^\perp = \left\{ \begin{bmatrix} 0 & z \\ -\bar{z}^t & 0 \end{bmatrix} \mid z = \begin{bmatrix} z_1 \\ z_2 \end{bmatrix}, z_i \in \mathbb{C} \right\},$$

one easily checks that $[\mathfrak{k}, \mathfrak{k}^\perp] \subset \mathfrak{k}^\perp$, so that $\operatorname{Ad}_K(\mathfrak{k}^\perp) \subset \mathfrak{k}^\perp$. It follows that $\langle , \rangle_\phi$ is Ad_K-invariant. Notice also that

$$[\mathfrak{k}^\perp, \mathfrak{k}^\perp] \subset \mathfrak{k}. \tag{2.5.2}$$

For nonzero k, $l \in \mathbb{R}$, consider the circle subgroup

$$H_{k,l} = \left\{ \begin{bmatrix} e^{2\pi ikt} & 0 & 0 \\ 0 & e^{2\pi ilt} & 0 \\ 0 & 0 & e^{-2\pi i(k+l)t} \end{bmatrix} \mid t \in \mathbb{R} \right\}. \tag{2.5.3}$$

The family of Aloff-Wallach examples consists of the seven-dimensional manifolds $M_{k,l} := G/H_{k,l}$ together with the (nonnormal) G-homogeneous metric for which the projection $\pi : G_\phi \to M_{k,l}$ becomes a Riemannian submersion (notice that $H_{k,l} \subset K$, so that $\langle , \rangle_\phi$ is right-invariant under $H_{k,l}$, and we may apply Theorem 2.4.1). For simplicity of notation, from now on, $\langle , \rangle_\phi$ will be denoted by $\langle , \rangle$ unless explicitly stated otherwise. Before computing the curvature of $M_{k,l}$, some preliminaries are in order.

The vertical space at the identity is the Lie algebra $\mathfrak{h}$ of $H_{k,l}$ which is spanned by

$$T = \begin{bmatrix} 2\pi ik & 0 & 0 \\ 0 & 2\pi il & 0 \\ 0 & 0 & -2\pi i(k+l) \end{bmatrix}.$$

The horizontal space $\mathcal{H}_e = \mathfrak{h}^\perp$ decomposes as an orthogonal direct sum $\mathcal{H}_e = H_1 \oplus H_2$, where $H_1 := \mathfrak{h}^\perp \cap \mathfrak{k}$, $H_2 := \mathfrak{k}^\perp$. The bracket relations between these spaces are as follows:

Lemma 2.5.1.

1. $[\mathfrak{h}, H_i] \subset H_i$;
2. $[H_i, H_i] \subset \mathfrak{h} \oplus H_1$;
3. $[H_1, H_2] \subset H_2$.

Proof. (1): Let $T_i \in \mathfrak{h}$, $i = 1, 2$. Since ad_{T_1} is skew-adjoint,

$$\langle [T_1, X], T_2 \rangle = -\langle X, [T_1, T_2] \rangle = 0$$

for $X \in \mathfrak{h}^\perp$, and $[\mathfrak{h}, \mathfrak{h}^\perp] \subset \mathfrak{h}^\perp$. Together with $[\mathfrak{h}, H_1] \subset \mathfrak{k}$ (which follows from the fact that $\mathfrak{h}, H_1 \subset \mathfrak{k}$), this means that $[\mathfrak{h}, H_1] \subset \mathfrak{h}^\perp \cap \mathfrak{k} = H_1$. On the other hand,

$$[\mathfrak{h}, H_2] = [\mathfrak{h}, \mathfrak{k}^\perp] \subset [\mathfrak{k}, \mathfrak{k}^\perp] \subset \mathfrak{k}^\perp = H_2.$$

(2): Since $\mathfrak{k} = \mathfrak{h} \oplus H_1$ and $H_1 \subset \mathfrak{k}$, $[H_1, H_1] \subset \mathfrak{k} = \mathfrak{h} \oplus H_1$. The other identity $[H_2, H_2] \subset \mathfrak{h} \oplus H_1$ is just (2.5.2).

(3): $[H_1, H_2] \subset [\mathfrak{k}, \mathfrak{k}^\perp] \subset \mathfrak{k}^\perp = H_2$. □

Lemma 2.5.2. *Suppose $kl > 0$, and $[X, Y] = 0$ for X, $Y \in \mathcal{H}_e = H_1 \oplus H_2$.*

1. *If X, $Y \in H_1$, or if X, $Y \in H_2$, then they are linearly dependent.*
2. *If $X \in H_1$ and $Y \in H_2$, then X or Y is zero.*

Proof. (1): Let

$$X = \begin{bmatrix} \alpha_1 i & z_1 & \\ -\bar{z}_1 & \beta_1 i & \\ & & -(\alpha_1+\beta_1)i \end{bmatrix}, \quad Y = \begin{bmatrix} \alpha_2 i & z_2 & \\ -\bar{z}_2 & \beta_2 i & \\ & & -(\alpha_2+\beta_2)i \end{bmatrix} \in H_1.$$

Then

$$[X,Y] = \begin{bmatrix} 2\operatorname{Im}\bar{z}_1 z_2 & iz_1(\beta_2-\alpha_2)+iz_2(\alpha_1-\beta_1) & 0 \\ i\bar{z}_1(\beta_2-\alpha_2)+i\bar{z}_2(\alpha_1-\beta_1) & -2\operatorname{Im}\bar{z}_1 z_2 & 0 \\ 0 & 0 & 0 \end{bmatrix}$$

is zero by assumption. Furthermore, the condition X, $Y \perp T$ implies

$$\left\langle \begin{bmatrix} \alpha_i \\ \beta_i \end{bmatrix}, \begin{bmatrix} 2k+l \\ 2l+k \end{bmatrix} \right\rangle = 0, \qquad i = 1,2. \tag{2.5.4}$$

Thus, we may assume

$$\begin{bmatrix} \alpha_1 \\ \beta_1 \end{bmatrix} = \lambda \begin{bmatrix} \alpha_2 \\ \beta_2 \end{bmatrix}, \qquad \lambda \in \mathbb{R}, \tag{2.5.5}$$

and the $(1,2)$ entry of $[X,Y]$ reads

$$(\lambda i z_2 - i z_1)(\alpha_2 - \beta_2) = 0. \tag{2.5.6}$$

If $\alpha_2 = \beta_2 = 0$, then $\alpha_1 = \beta_1 = 0$ by (2.5.5); moreover, z_1, z_2 are linearly dependent since $\operatorname{Im}\bar{z}_1 z_2 = 0$, and the claim follows. If $\alpha_2 = \beta_2 \neq 0$, then by (2.5.4), $3(k+l) = 0$, contrary to assumption. Finally, if $\alpha_2 \neq \beta_2$, then by (2.5.6), $z_1 = \lambda z_2$, and together with (2.5.5), we obtain $X = \lambda Y$ as claimed.

Suppose next that X, $Y \in H_2$. Since H_2 is the horizontal space for the submersion $SU(3) \to \mathbb{C}P^2$, $R(\pi_* X, \pi_* Y) = 0$ by (2.4.4). But $\mathbb{C}P^2$ has positive curvature, so X and Y must be linearly dependent.

(2): Let

$$X = \begin{bmatrix} \alpha i & w & \\ -\bar{w} & \beta i & \\ & & -(\alpha+\beta)i \end{bmatrix} \in H_1, \quad Y = \begin{bmatrix} 0 & z \\ -\bar{z}^t & 0 \end{bmatrix} \in H_2, \quad z = \begin{bmatrix} z_1 \\ z_2 \end{bmatrix} \in \mathbb{C}^2.$$

The assumption $X \perp T$ implies $(2\alpha+\beta)k + (2\beta+\alpha)l = 0$, which together with the fact that $kl > 0$ yields

$$2\alpha + \beta = -\frac{l}{k}(2\beta + \alpha). \tag{2.5.7}$$

Since

$$0 = [X,Y] = \begin{bmatrix} 0 & 0 & (2\alpha+\beta)iz_1 + wz_2 \\ 0 & 0 & (\alpha+2\beta)iz_2 - \bar{w}z_1 \\ * & * & 0 \end{bmatrix},$$

(2.5.7) yields the system

$$-\frac{l}{k}(2\beta+\alpha)iz_1 + wz_2 = 0,$$
$$-\bar{w}z_1 + (2\beta+\alpha)iz_2 = 0$$

of equations in z_1, z_2, with determinant of coefficients $(l/k)(2\beta+\alpha)^2 + |w|^2$. But $kl > 0$, so this determinant is positive (implying $z = 0$, and thus, $Y = 0$) unless $2\beta = -\alpha$ and $w = 0$. In the latter case, (2.5.7) implies $\alpha = \beta = w = 0$; i.e., $X = 0$. □

For $X \in \mathcal{H}_e$, write $X = X_1 + X_2 \in H_1 \oplus H_2$.

Lemma 2.5.3. *If $kl > 0$ and $[X, Y] = [X_1, Y_1] = 0$, then X and Y are linearly dependent.*

Proof. Consider first the case $X_1 = 0$ and $X_2 \neq 0$. Then $0 = [X, Y] = [X_2, Y_2] + [X_2, Y_1]$. Since $[X_2, Y_2] \in \mathfrak{h} \oplus H_1$, and $[X_2, Y_1] \in H_2$ by Lemma 2.5.1, they both vanish. By Lemma 2.5.2 (2), $Y_1 = 0$, so that $0 = [X, Y] = [X_2, Y_2]$. Lemma 2.5.2 (1) then implies that $X_2 = X$ and $Y_2 = Y$ are linearly dependent. The case $X_1 \neq 0$ and $X_2 = 0$ is similar. We may therefore assume that X_1 and X_2 are both nonzero. Then

$$0 = [X, Y] = [X_1, Y_2] + [X_2, Y_1] + [X_2, Y_2], \tag{2.5.8}$$

and by Lemma 2.5.1, $[X_2, Y_2] = 0$. Lemma 2.5.2 (1) then implies that $Y_2 = \alpha X_2$ for some $\alpha \in \mathbb{R}$, and therefore $Y_1 = \beta X_1$, also for some $\beta \in \mathbb{R}$. Substituting in (2.5.8) yields

$$0 = [X_1, Y_2] + [X_2, Y_1] = (\alpha - \beta)[X_1, X_2].$$

But $[X_1, X_2] \neq 0$ by Lemma 2.5.2 (2), since $X_i \neq 0$. Thus, $\alpha = \beta$ and $Y = \alpha X$ as claimed. □

We are now ready to prove the main result:

Theorem 2.5.1. *Let $H_{k,l}$ denote the circle group from (2.5.3). If $kl(k+l) > 0$, then $M_{k,l} := SU(3)/H_{k,l}$ admits a homogeneous metric of positive curvature.*

Proof. Endow $SU(3)$ with the warped metric $\langle , \rangle_\phi$ the curvature of which is given by (2.5.1). As remarked earlier, this metric is right-invariant under $H_{k,l}$, and by Theorem 2.4.1, $M_{k,l}$ admits a metric for which the projection $\pi : SU(3) \to M_{k,l}$ becomes a Riemannian submersion. Given X, Y in the horizontal space $\mathcal{H}_e$ of $SU(3)$ at the identity e, a lengthy but straightforward computation using (2.5.1), Lemma 2.5.1 and the results from Section 2.5 implies that the curvature tensor R_M of $M_{k,l}$ is given by

$$\begin{aligned}\langle R_M(\pi_* X, \pi_* Y)\pi_* Y, \pi_* X\rangle = {} & \frac{1-3t}{4}|[X,Y]_1|^2 + t^2|[X_1,Y_1]|^2 \\ & + \frac{(1+t)^2}{4}|[X,Y]_2|^2 + |[X,Y]^{\mathrm{v}}|^2 \\ & + (t - t^2)\langle [X_1, Y_1], [X, Y]\rangle.\end{aligned}$$

Consider the quadratic forms

$$Q_1(x,y) = \frac{1-3t}{4}x^2 - |t-t^2|xy + t^2y^2,$$
$$Q_2(x,y) = x^2 - |t-t^2|xy + t^2y^2,$$

which are easily seen to be positive definite if $t \in (-1,0)$. By the Cauchy-Schwartz inequality,

$$\begin{aligned}\langle R_M(\pi_*X,\pi_*Y)\pi_*Y,\pi_*X\rangle \geq{}& \frac{1-3t}{4}|[X,Y]_1|^2 - |t-t^2||[X,Y]_1||[X_1,Y_1]_1| \\ &+ t^2|[X_1,Y_1]_1|^2 + |[X,Y]^{\mathsf{v}}|^2 \\ &- |t-t^2||[X,Y]^{\mathsf{v}}||[X_1,Y_1]^{\mathsf{v}}| + t^2|[X_1,Y_1]^{\mathsf{v}}|^2 \\ &+ \frac{(1+t)^2}{4}|[X,Y]_2|^2 \\ ={}& Q_1(|[X,Y]_1|,|[X_1,Y_1]_1|) \\ &+ Q_2(|[X,Y]^{\mathsf{v}}|,|[X_1,Y_1]^{\mathsf{v}}|) \\ &+ \frac{(1+t)^2}{4}|[X,Y]_2|^2 \\ \geq{}& 0\end{aligned}$$

if $t \in (-1,0)$. Furthermore, it can only be zero if

$$|[X,Y]_1| = |[X,Y]_2| = |[X,Y]^{\mathsf{v}}| = |[X_1,Y_1]_1| = |[X_1,Y_1]^{\mathsf{v}}| = 0.$$

In this case, $[X,Y] = [X_1,Y_1] = 0$, and by Lemma 2.5.3, X and Y are linearly dependent. Thus, $M_{k,l}$ has positive curvature if $-1 < t < 0$. □

It can be shown that if k, l are relatively prime, then $H^4(M_{k,l}) = \mathbb{Z}/r\mathbb{Z}$, where $r = |k^2 + l^2 + kl|$. The above result therefore provides an infinite family of seven-dimensional homogeneous spaces of positive curvature. These, together with the rank one symmetric spaces and five exceptional manifolds are known to be the only simply connected homogeneous spaces of positive curvature, cf. [20], [16], and [129].

2.6 Bi-quotients of Lie groups

Consider a Lie group G with a left-invariant metric that is right-invariant under a subgroup H. The last remark in the previous section implies that in order to obtain more positively curved manifolds from G and H, one must somehow generalize the homogeneous space construction G/H. Now, observe that $G \times H$ acts isometrically on G via

$$\begin{aligned}(G\times H)\times G &\longrightarrow G,\\ ((g,h),a) &\longmapsto (g,h)a := gah^{-1},\end{aligned}$$

and so does any subgroup K of $G\times H$. If K acts freely on G, then the space $G//K$ of orbits is called a *bi-quotient* of G. Since the action of K is by isometries, there is a natural Riemannian metric on the quotient space such that the projection $\pi : G \to G//K$ becomes a Riemannian submersion. Notice that if $K \subset \{e\} \times H$, then $G//K$ is just a homogeneous space in the sense of Section 2.5. In general, though, the vertical distribution of G is not invariant under left translation: the fiber through $g \in G$ is $F^g = \{(k_1 g k_2^{-1}) \mid (k_1, k_2) \in K\}$. Given $U = (U_1, U_2) \in \mathfrak{k}$, let c_i denote the one-parameter subgroup of G with $\dot{c}_i(0) = U_i(e)$. Then $t \mapsto c(t) := c_1(t) g c_2^{-1}(t)$ is a curve in F^g, and U determines a vertical Killing field $\tilde{U}$, with

$$\tilde{U}(g) := \dot{c}(0) = (R_{g*}U_1 - L_{g*}U_2)(e). \tag{2.6.1}$$

The vertical space $\mathcal{V}_g$ at g is then equal to the subspace spanned by all $\tilde{U}(g)$, as U ranges over $\mathfrak{k}$. Since left translation is an isometry, we may, as far as calculations are concerned, translate this space back to the origin and use (2.6.1) to obtain

$$\mathcal{V}_e^g := L_{g^{-1}*}\mathcal{V}_g = \operatorname{span}\{(\operatorname{Ad}_{g^{-1}} U_1 - U_2)(e) \mid (U_1, U_2) \in \mathfrak{k}\}. \tag{2.6.2}$$

We see from this that the subspaces $\mathcal{V}_e^g$ of G_e need not coincide for different values of g.

Bi-quotients have been used to construct examples of spaces with positive or nonnegative curvature. We discuss two such here.

2.6.1 The Gromoll-Meyer exotic sphere

Recall that the *symplectic group* $Sp(n)$ is the group of $n \times n$ quaternion matrices A satisfying $A\bar{A}^t = I_n$. If we identify $\mathbb{R}^4$ with the division algebra $\mathbb{H}$ of quaternions, then $Sp(1)$ is just the set of unit quaternions, or in other words, S^3. $Sp(n)$ is simple, and as such, admits a canonical bi-invariant metric of nonnegative curvature by Proposition 2.4.2. Set $G = Sp(2)$, and consider the subgroup K of $G \times G$ given by

$$K = \left\{\left(\begin{bmatrix} q & 0 \\ 0 & q \end{bmatrix}, \begin{bmatrix} \bar{q} & 0 \\ 0 & 1 \end{bmatrix}\right) \mid q \in Sp(1)\right\}.$$

K acts freely on G, so that the seven-dimensional bi-quotient $M := G//K$ of G admits a metric with nonnegative sectional curvature. It turns out that this metric has positive curvature on an open set.

Before identifying M as an exotic sphere (i.e., as a manifold homeomorphic, but not diffeomorphic, to a standard sphere), we recall part of Milnor's description of S^3-bundles over S^4 [86]:

Denote by x_N and x_S the stereographic projections of S^4 from the north and south poles respectively onto $\mathbb{R}^4$. For $u \in \mathbb{H} \setminus \{0\}$,

$$x_N \circ x_S^{-1}(u) = x_S \circ x_N^{-1}(u) = \frac{u}{|u|^2},$$

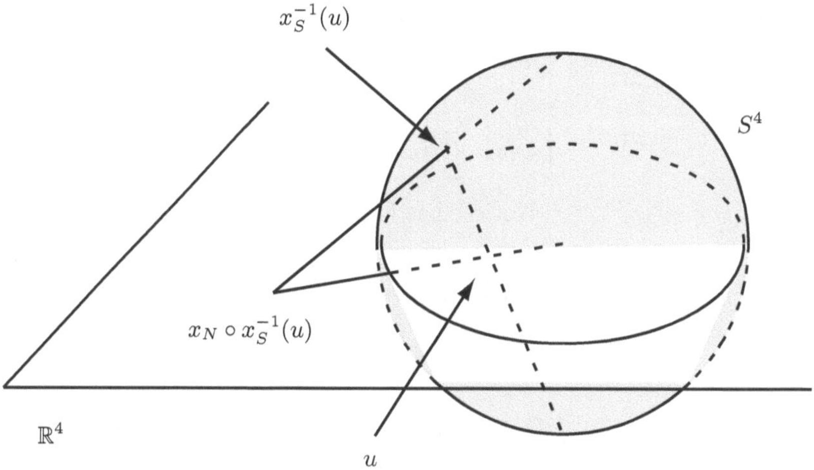

so that S^4 is the identification space

$$\mathbb{R}^4 \sqcup \mathbb{R}^4/\sim, \qquad u \sim \frac{u}{|u|^2}, \quad u \neq 0,$$

obtained as the disjoint union of two copies of $\mathbb{R}^4$, where each nonzero u in one copy is identified with $u/|u|^2$ in the other. Next, let l be an odd integer, and set $m := (1+l)/2$, $n := (1-l)/2$. Define

$$E_l = \mathbb{R}^4 \times S^3 \sqcup \mathbb{R}^4 \times S^3/\sim,$$

where (u,q) (for $u \neq 0$) in the first copy is identified with $(u/|u|^2, u^m q u^n/|u|)$ in the second copy. Multiplication here is understood to be quaternion multiplication. It follows that there exists a well-defined map $\pi : E_l \to S^4$, with $\pi[u,q] = x_N^{-1}(u)$ if (u,q) lies in the first copy, and $\pi[u,q] = x_S^{-1}(u)$ for (u,q) in the second one (here, $[u,v]$ denotes the equivalence class of (u,v) in E_l). Define maps $\phi_N : \mathbb{R}^4 \times S^3 \sqcup \emptyset/\sim \to S^3$ and $\phi_S : \emptyset \sqcup \mathbb{R}^4 \times S^3/\sim \to S^3$ by $\phi_N[u,q] = q$, $\phi_S[u,q] = q$. Then $\mathcal{A} = \{(x_N \circ \pi, \phi_N), (x_S \circ \pi, \phi_S)\}$ determines a topology on E_l by requiring both maps to be homeomorphisms. Since

$$(x_S \circ \pi, \phi_S) \circ (x_N \circ \pi, \phi_N)^{-1}(u,q) = \left(\frac{u}{|u|^2}, \frac{u^m q u^n}{|u|}\right), \qquad u \neq 0,$$

the atlas $\mathcal{A}$ determines a differentiable structure on E_l, and at the same time an S^3-bundle structure on $\pi : E_l \to S^4$. Milnor showed that E_l is homeomorphic to S^7, but not diffeomorphic, unless l^2 is congruent to 1 modulo 7. We now identify M as an exotic sphere, by showing it is diffeomorphic to E_3: Denote by $\rho : Sp(2) \to M$

the projection onto the orbit space, and define

$$f_1 : \mathbb{R}^4 \times S^3 \to M, \qquad f_2 : \mathbb{R}^4 \times S^3 \to M,$$

$$(u,q) \mapsto \rho\left(\phi(u)\begin{bmatrix} q & \bar{u} \\ -uq & 1 \end{bmatrix}\right) \qquad (v,p) \mapsto \rho\left(\phi(v)\begin{bmatrix} \bar{v}p & 1 \\ -\bar{p} & v \end{bmatrix}\right)$$

where $\phi(u) = (1+|u|^2)^{-1/2}$. Notice that b and d cannot both vanish for

$$\rho\left(\begin{bmatrix} a & b \\ c & d \end{bmatrix}\right) \in M.$$

If $d \neq 0$, then

$$\begin{aligned} \rho\left(\begin{bmatrix} a & b \\ c & d \end{bmatrix}\right) &= \rho\left(\frac{\bar{d}}{|d|}\begin{bmatrix} a & b \\ c & d \end{bmatrix}\begin{bmatrix} \frac{d}{|d|} & 0 \\ 0 & 1 \end{bmatrix}\right) = \rho\left(|d|\begin{bmatrix} \frac{\bar{d}ad}{|d|^3} & \frac{\bar{d}b}{|d|^2} \\ \frac{\bar{d}cd}{|d|^3} & 1 \end{bmatrix}\right) \\ &= f_1\left(\frac{\bar{b}d}{|d|^2}, \frac{\bar{d}ad}{|d|^3}\right). \end{aligned}$$

Similarly, if $b \neq 0$, then

$$\rho\left(\begin{bmatrix} a & b \\ c & d \end{bmatrix}\right) = f_2\left(\frac{\bar{b}d}{|b|^2}, -\frac{\bar{b}cb}{|b|^3}\right).$$

It follows that the union of the images of f_1 and f_2 equals all of M. Moreover, f_1 and f_2 are differentiable, and by the above, are invertible with

$$f_1^{-1}\left(\rho\begin{bmatrix} a & b \\ c & d \end{bmatrix}\right) = \left(\frac{\bar{b}d}{|d|^2}, \frac{\bar{d}ad}{|d|^3}\right), \qquad f_2^{-1}\left(\rho\begin{bmatrix} a & b \\ c & d \end{bmatrix}\right) = \left(\frac{\bar{b}d}{|b|^2}, -\frac{\bar{b}cb}{|b|^3}\right).$$

Finally, $f_2^{-1} \circ f_1(u,q) = (u/|u|^2, u^2qu^{-1}/|u|)$, so that f_1 and f_2 combine to yield a diffeomorphism between E_3 and M, as claimed; see also [59].

2.6.2 The seven-dimensional Eschenburg examples

Let $G = SU(3)$, and denote by $M_{k,l} = G/H_{k,l}$ any one of the Aloff-Wallach spaces of positive curvature from Section 2.5. Eschenburg constructed a sequence M_i of bi-quotients of G with curvature converging in a sense to that of $M_{k,l}$ [42], [43]. In particular, M_i is positively curved for large i. One remarkable aspect is that M_i is not a Riemannian homogeneous space; in fact, he proves that it is not even homotopy equivalent to a compact homogeneous space.

For $n \in \mathbb{N}$, consider the subgroup K_n of $G \times G$ given by

$$K_n = \left\{\left(\begin{bmatrix} e^{2\pi it} & 0 & 0 \\ 0 & 1 & 0 \\ 0 & 0 & e^{-2\pi it} \end{bmatrix}, \begin{bmatrix} e^{2\pi inkt} & 0 & 0 \\ 0 & e^{2\pi inlt} & 0 \\ 0 & 0 & e^{-2\pi in(k+l)t} \end{bmatrix}\right) \mid t \in \mathbb{R}\right\}.$$

For the sake of brevity, we denote a diagonal matrix such as the first one above by $\mathrm{diag}(e^{2\pi i t}, 1, e^{-2\pi i t})$. K_n acts by isometries on G via

$$((a,b),g) \mapsto agb^{-1},$$

and it is readily checked that this action is free for infinitely many values of n. We will denote by M_n the bi-quotient $G//K_n$ of G.

The Lie algebra $\mathfrak{k}_n$ of K_n is spanned by $(U_n, V) \in \mathfrak{g} \times \mathfrak{g}$, where

$$U_n = \mathrm{diag}\left(\frac{2\pi i}{n}, 0, -\frac{2\pi i}{n}\right), \qquad V = \mathrm{diag}(2\pi i k, 2\pi i l, -2\pi i(k+l)).$$

Notice that V generates the Lie algebra of $H_{k,l} \subset G$, where $G/H_{k,l}$ is the Aloff-Wallach space $M_{k,l}$. The following result then guarantees that M_n has positive curvature for infinitely many values of n:

Theorem 2.6.1. *Let K_i, $i = 1, 2, \dots$ be 1-parameter subgroups of $G \times G$ acting freely on G, with Lie algebras $\mathfrak{k}_{n_i}$ spanned by (U_i, V), where $U_i \to 0$, $V \neq 0$. Denote by H the subgroup of G with Lie algebra spanned by V. If $M := G/H$ has positive curvature, then so does $M_i := G//K_i$ for large enough i.*

Proof. Set $T_i^g := (V - \mathrm{Ad}_{g^{-1}} U_i)(e)/|(V - \mathrm{Ad}_{g^{-1}} U_i)(e)| \in G_e$, for $g \in G$. Then

$$T_i^g \to \frac{V}{|V|}(e) \tag{2.6.3}$$

uniformly in g. Fix some such g, and denote by $\mathcal{H}_g$, $\mathcal{H}_g^i \subset G_g$ the horizontal spaces of the submersions $\pi : G \to M := G/H$, $\pi_i : G \to M_i = G//K_i$, respectively. Observe that $\pi_{i*|\mathcal{H}_g}$ has maximal rank for large i by (2.6.2) and (2.6.3). Next, let S_g be the unit sphere in $\mathcal{H}_g$, $h : S_g \to G_g$ the inclusion, and $h_i : S_g \to G_g$ the restriction of the π_i-horizontal projection $G_g \to \mathcal{H}_g^i$ to S_g, followed by inclusion $\mathcal{H}_g^i \hookrightarrow G_g$. Then $h_i \to h$ uniformly, since

$$h_i(x) = L_{g*}\big(L_{g^{-1}*}x - \langle L_{g^{-1}*}x, T_i^g\rangle T_i^g\big), \qquad x \in S_g.$$

By Theorem 1.5.1, given a plane $P \subset \mathcal{H}_g$ spanned by π-horizontal fields X, Y that are orthonormal at g, and $P_i := \mathrm{span}\{h_i X(g), h_i Y(g)\}$,

$$\begin{aligned} K_{M_i}(\pi_{i*}P) &= K_G(P_i) + \frac{3|(1_{G_g} - h_i)[h_i X(g), h_i Y(g)]|^2}{4(|h_i X(g)|^2 |h_i Y(g)|^2 - \langle h_i X(g), h_i Y(g)\rangle^2)} \\ &\to K_G(P) + \frac{3}{4}|(1_{G_g} - h)[X,Y](g)|^2 = K_M(\pi_* P). \end{aligned}$$

The claim now follows by compactness of G. ☐

Using spectral sequences, it can be shown that there exists a sequence $n_i \to \infty$ of positive integers such that the spaces $M_{n_i} = G//K_{n_i}$ are *strongly inhomogeneous*; i.e., they are not homotopy equivalent to any compact Riemannian homogeneous space. The above construction therefore yields infinite families of positively curved such spaces.

Just as the Eschenburg examples are derived from the Aloff-Wallach spaces, Bazaikin, motivated by the Berger example of the normal homogeneous space $SU(5)/(Sp(2) \times U(1))$, discovered an infinite family of thirteen-dimensional positively curved manifolds [12]. It is known that each Bazaikin space contains one or more totally geodesically imbedded Aloff-Wallach or Eschenburg spaces [39].

2.7 Associated bundles

There is yet another construction which yields a large class of homogeneous submersions that often appear in the literature. Let P denote the total space of a (right) principal G-bundle $\pi_P : P \to M = P/G$, and F a manifold on which G acts (on the left). Then G acts freely on $P \times F$ via:

$$\begin{aligned} G \times P \times F &\to P \times F, \\ (g, p, m) &\mapsto (pg^{-1}, gm). \end{aligned}$$

The quotient manifold $G\backslash(P \times F)$ is usually denoted by $P \times_G F$, and is the total space of a bundle $\pi : P \times_G F \to M$ with fiber F and group G, where $\pi(G(p, m)) = \pi_P(p)$ for $p \in P$, $m \in F$. π is called *the bundle with fiber F associated to the principal bundle π_P* (and the given action of G on F). Now, suppose that both P and F have G-invariant metrics. Then G acts by isometries on the Riemannian product $P \times F$, and by Examples and Remarks 1.2.1 (iv), $P \times_G F$ inherits a metric such that the projection $\rho : P \times F \to P \times_G F$ is a Riemannian submersion. Similarly, there exists a metric on M for which $\pi_P : P \to M$ becomes Riemannian. Let $\pi_1 : P \times F \to P$ denote the projection onto the first factor. Since the diagram

$$\begin{array}{ccc} P \times F & \xrightarrow{\ \rho\ } & P \times_G F \\ {\scriptstyle \pi_1}\downarrow & & \downarrow{\scriptstyle \pi} \\ P & \xrightarrow[\pi_P]{} & M \end{array}$$

commutes, and since π_1, ρ, and π_P are Riemannian, the bundle projection $\pi : P \times_G F \to M$ is also a Riemannian submersion. Furthermore, if π_P has totally geodesic fibers, then so does π. To see this, consider a vertical vector u in the tangent space of $P \times_G F$ at $\rho(p, m)$, and its ρ-horizontal lift $\tilde{u} = (v, w) \in P_p \times F_m$. The geodesic $c_{\tilde{u}}$ in $P \times F$ with $\dot{c}_{\tilde{u}}(0) = \tilde{u}$ decomposes as a pair (c_1, c_2) of geodesics in each factor, and $c := \rho \circ c_{\tilde{u}}$ is a geodesic in $P \times_G F$ with $\dot{c}(0) = u$. It suffices to show that $\pi_* \dot{c} \equiv 0$. Now, $\pi_* u = 0$, so that

$$\pi_{P*}(v) = \pi_{P*} \circ \pi_{1*}(\tilde{u}) = \pi_* \circ \rho_*(\tilde{u}) = \pi_* u = 0.$$

Since π_P has totally geodesic fibers, $\pi_{P*}(\dot{c}_1) \equiv 0$. But then,

$$\pi_* \dot{c} = \pi_* \circ \rho_*(\dot{c}_{\tilde{u}}) = \pi_{P*} \circ \pi_{1*}(\dot{c}_{\tilde{u}}) = \pi_P(\dot{c}_1) \equiv 0,$$

which establishes the claim, cf. also [96].

Example 2.7.1 (The tangent bundle of a homogeneous space). Let $M = G/H$ denote a Riemannian homogeneous space, and set $p := eH \in M$. There are natural actions $(g, aH) \mapsto gaH$ and $(g, u) \mapsto g_* u$ of G on M and TM, respectively. The latter action, when restricted to H, leaves the tangent space of M at p invariant, and thus defines a bundle $\pi : G \times_H M_p \to G/H$ with fiber M_p associated to the principal H-bundle $G \to G/H$. G acts on the total space $G \times_H M$ via $(g_1, \rho(g_2, u)) \mapsto \rho(g_1 g_2, u)$, and the map

$$\begin{aligned} f : G \times_H M_p &\to TM, \\ \rho(g, u) &\mapsto g_* u \end{aligned}$$

is a well-defined G-equivariant diffeomorphism. Since the action of H on both factors is by isometries (here M_p is identified with Euclidean space by means of its inner product and H is then a subgroup of the orthogonal group), $G \times_H M_p$ inherits a natural Riemannian metric, and so does TM via f. Furthermore, if $\pi_M : TM \to M$ denotes the vector bundle projection that maps $v \in M_q$ to q, then $\pi_M \circ f = \pi$, since for $u \in M_p$,

$$\pi_M \circ f(\rho(g, u)) = \pi_M(g_* u) = g(\pi_M u) = g(p) = \pi(\rho(g, u)).$$

Thus, after identifying $G \times_H M_p$ with TM via f, π is just the vector bundle projection π_M.

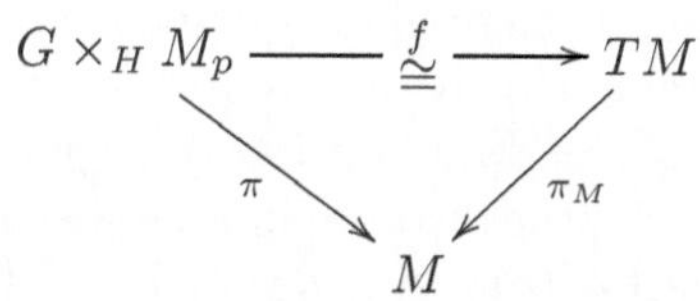

By Theorem 2.4.1 together with the above discussion, $\pi : TM \to M$ is then a Riemannian submersion with totally geodesic fibers. Notice that if the metric on G is bi-invariant, then the sectional curvature of TM is nonnegative. This is the case, for example, of the tangent bundle of $S^n = SO(n+1)/SO(n)$. The curvature cannot, however, be positive: Since $M = G \times_H \{0\} \subset G \times_H M_p$, the horizontal distribution through any point of M is integrable; i.e., $A \equiv 0$ along M, or equivalently, the zero section is horizontal. By the proof of Theorem 2.2.2, the S-tensor must also vanish along M, and any plane spanned by a vector tangent to M and one orthogonal to M has zero curvature, cf. (1.5.2).

The construction above has many applications. One such is the following theorem due to Cheeger [34]:

Theorem 2.7.1. *The connected sum of two rank one symmetric spaces of the same dimension admits a metric of nonnegative curvature.*

Proof. The construction is similar in all cases, and we shall only outline it for complex projective space. Consider the Hopf fibration $S^{2n+1} \to \mathbb{C}P^n$; the associated rank two vector bundle $E = S^{2n+1} \times_{S^1} \mathbb{R}^2 \to \mathbb{C}P^n$ is the normal bundle of the inclusion of $\mathbb{C}P^n$ into $\mathbb{C}P^{n+1}$, and as such, its total space E is diffeomorphic to $\mathbb{C}P^{n+1}$ with a ball removed. Endow $\mathbb{R}^2 \setminus \{0\}$ with the metric given in polar coordinates by

$$\langle \partial_r, \partial_r \rangle \equiv 1, \quad \langle \partial_r, \partial_\theta \rangle \equiv 0, \quad \text{and } \langle \partial_\theta, \partial_\theta \rangle(r) = f^2(r),$$

where f is a differentiable concave function on $(0, +\infty)$ with the following properties:

1. f is extendable to a smooth odd function on $(-\epsilon, \epsilon)$ for some $\epsilon > 0$;
2. $f'(0) = 1$, and
3. f is constant for $r \geq r_0$, where $r_0 > 0$.

Conditions (1) and (2) guarantee that this metric on the punctured plane is extendable to all of $\mathbb{R}^2$, cf. [53], [104]. The resulting surface P_2 has nonnegative curvature. In fact, it can be expressed as a warped product

$$P_2 = [0, \infty) \times_{f^2} S^1,$$

where by Corollary 2.2.1,

$$\langle R(X,T)T, X \rangle = -\frac{f''}{f} |T|^2 \geq 0.$$

Notice that the warping function f is constant for $r \geq r_0$, so that the complement of the ball of radius r_0 around the origin is isometric to the cylinder $(r_0, \infty) \times S^1$. Since P_2 is rotationally symmetric, there is a submersion metric on $E = S^{2n+1} \times_{S^1} P_2$ induced by $\rho : S^{2n+1} \times P_2 \to E$, and it has nonnegative sectional curvature.

Given $p \in S^{2n+1}$, $u \in \mathbb{R}^2$, the distance in E between $\rho(p, 0)$ and $\rho(p, u)$ equals the distance between the sets $\{(pz_0, 0) \mid z_0 \in S^1\}$ and $\{(pz_1, z_1^{-1}u) \mid z_1 \in S^1\}$ in the Riemannian product $S^{2n+1} \times P_2$. The latter is clearly $|u|$, and it follows that the distance from $\rho(p, u)$ to the zero section $\rho(S^{2n+1} \times \{0\}) = \mathbb{C}P^n$ also equals $|u|$. Since P_2 is isometric to the product $(r_0, \infty) \times S^1$ for $r \geq r_0$, the action of S^1 on the first factor is trivial, and the complement of the tubular neighborhood of radius r_0 about $\mathbb{C}P^n$ is isometric to

$$S^{2n+1} \times_{S^1} (S^1 \times (r_0, \infty)) = (S^{2n+1} \times_{S^1} S^1) \times (r_0, \infty) = M \times (r_0, \infty),$$

where M denotes the Riemannian manifold $S^{2n+1} \times_{S^1} S^1$ which is diffeomorphic to a sphere.

By gluing two such disk bundles of radius $R > r_0$ along their common boundary, we then obtain a well-defined metric of nonnegative curvature on the connected sum of two copies of $\mathbb{C}P^{n+1}$. □

Remark 2.7.1. One cannot expect to construct metrics of nonnegative curvature on a connected sum of arbitrarily many copies of $\mathbb{C}P^n$: One of the known obstructions to the existence of metrics with nonnegative curvature is a result of Gromov [64], which states that for each positive integer n, there exists a constant $C(n)$, such that the total Betti number of any n-dimensional complete manifold of nonnegative sectional curvature is less than $C(n)$. This, in fact, illustrates how much weaker, as a property, positive Ricci curvature is as opposed to nonnegative sectional curvature: Sha and Yang [113] have shown the existence of seven-dimensional manifolds with arbitrarily large total Betti number, which admit complete metrics of positive Ricci curvature (and cannot all admit metrics of nonnegative sectional curvature by the above). The details of their construction are fairly technical in nature, but since it is related to the metric on the Hopf bundle from the above theorem, we provide an overview of it.

First, let us look at an alternative description of the standard metric on the total space $E = S^3 \times_{S^1} \mathbb{R}^2$ of the plane bundle associated to the lowest-dimensional Hopf fibration. The unit sphere bundle has total space $S^3 \times_{S^1} S^1 = S^3$, so that the complement E_0 of the zero section $S^3 \times_{S^1} \{0\} = S^2$ is diffeomorphic to $(0,\infty) \times S^3$ via

$$\begin{aligned} E_0 &\to (0,\infty) \times S^3, \\ u &\mapsto (|u|, \frac{u}{|u|}). \end{aligned}$$

E itself can then be realized as $[0,\infty) \times S^3/\sim$, where the equivalence relation identifies $(0,p)$ with $(0,pz)$ for $p \in S^3$, $z \in S^1$. We have seen that the distance function from the zero section is the projection $\pi_1 : (0,\infty) \times S^3 \to (0,\infty)$, and is a Riemannian submersion. The total space $(0,\infty) \times S^3$ of the submersion is not, however, a warped product, because only the direction tangent to the Hopf fiber in S^3 is warped, leaving its orthogonal complement unchanged: Denote by I the Killing field on S^3 whose flow generates the Hopf fibers, cf. Examples and Remarks 1.5.1, where it was called IN instead. If $\mathbb{R}^2$ is endowed with the standard flat metric, then the vertical space of $\rho : S^3 \times \mathbb{R}^2 \to E = S^3 \times_{S^1} \mathbb{R}^2$ has $1/(1+r^2)^{1/2}(I, -\partial_\theta)$ as orthonormal basis. The vertical component $(I,0)^{\mathrm{v}}$ of $(I,0)$ then has norm squared

$$|(I,0)^{\mathrm{v}}|^2 = \frac{1}{1+r^2}\langle (I,0),(I,-\partial_\theta)\rangle^2 = \frac{1}{1+r^2},$$

so that $|\rho_*(I,0)|^2 = 1 - 1/(1+r^2) = r^2/(1+r^2)$. Thus, the metric on $E = [0,\infty) \times S^3/\sim$ is

$$dr^2 \oplus g_r,$$

where g_r is the metric on S^3 given by

$$g_r(I,I) = \frac{r^2}{1+r^2}, \qquad g_{r|I^\perp \otimes I^\perp} = g_{|I^\perp \otimes I^\perp},$$

with g denoting the canonical metric on S^3. As in the proof of Theorem 2.7.1, the function $r \mapsto r^2/(1+r^2)$ may be replaced by any function f^2, where f is odd, and $f(0) = 0$, $f'(0) = 1$. In the sequel, we shall consider the unit disk bundle $E^1 = [0,1] \times S^3/\sim$, together with the metric $dr^2 \oplus g_{f,b}$, where

$$g_{f,b}(I,I) = f^2, \qquad g_{f,b|I^\perp \otimes I^\perp} = b^2 g_{|I^\perp \otimes I^\perp}, \tag{2.7.1}$$

with the function f and the positive number b yet to be specified.

With these preliminaries out of the way, we now examine the Sha-Yang construction, which involves surgery on the seven-dimensional manifold $S^4 \times S^3$: Let $a > 1$, b as above, and consider the Euclidean spheres $M = S^4_a$, $N = S^3_b$ of radii a, b, respectively. Fix a point $p_0 \in M$. The closed ball $\bar{B}_1(p_0)$ of radius 1 around p_0 in M is isometric to the warped product

$$[0,1] \times_\phi S^3_1/\sim_0,$$

where $\phi(r) = a \sin(r/a)$, and $\sim_0$ collapses $\{0\} \times S^3_1$ to a point, cf. Theorem 2.2.1. The manifold $(M \setminus \bar{B}_1(p)) \times N$ is then $M \times N$ with

$$B_1(p) \times S^3_b = ([0,1] \times_\phi S^3_1)/\sim_0 \times S^3_b$$

removed. The part removed is now replaced by a space that is topologically

$$E^1 \times S^3 = ([0,1] \times S^3)/\sim \times S^3,$$

but with the identification interchanging the S^3-factors; i.e., $(r,p,q) \in B_1(p) \times S^3$ is to be identified with $(r,q,p) \in E^1 \times S^3$. The metric on $E^1 \times S^3$ is a warped product

$$(E^1, dr^2 \oplus g_{f,b}) \times_{k^2} S^3_1.$$

Because of the interchange in the S^3-factors, we require that

$$k(r) = a \sin \frac{r}{a}, \quad \text{and } f(r) = b \text{ for } r \geq 1,$$

in order to obtain a well-defined metric on the whole manifold.

Sha and Yang show that the functions k, f, and the number b may be chosen to yield a metric of positive Ricci curvature on $E^1 \times S^3$. Since the metric on the complement of $E^1 \times S^3$ is the original product metric, one obtains a metric of positive Ricci curvature on the entire manifold. Finally, by choosing the radius a of S^4 large enough, the same type of surgery can be performed at any given number of points on S^4, since they only need to be at distance larger than 2 from each other. One then obtains 7-manifolds of positive Ricci curvature with arbitrarily large total Betti numbers.

These examples have since been extended to four-dimensional manifolds, cf. [6], [114]: It is now known that a (clearly) necessary but also sufficient condition for a compact, simply connected four-dimensional manifold to admit a metric of positive Ricci curvature is that it be homeomorphic to one that admits a metric with positive scalar curvature. There are 4-manifolds that do not fall under this category, namely spin manifolds with nonzero signature.

Returning to our original theme, it should be noted that there is also a different way of constructing metrics on $P\times_G F$, one that does not involve a metric on P, but rather one on the base space M of the principal G-bundle $P \to M$, together with a connection on that bundle. The proposition below is due to Vilms [127]:

Proposition 2.7.1. *Let M be a Riemannian manifold, $\pi_P : P \to M$ a principal G-bundle over M with connection $\mathcal{H}$. Given a Riemannian manifold F on which G acts by isometries, there exists a (unique) metric on $P \times_G F$ such that $\pi : P \times_G F \to M$ is a Riemannian submersion with totally geodesic fibers isometric to F and horizontal distribution $\tilde{\mathcal{H}} := \rho_*(\mathcal{H} \times \{0\})$, where $\rho : P \times F \to P \times_G F$ denotes projection.*

Such a metric is called a *connection metric.*

Proof. As before, $\pi_1 : P \times F \longrightarrow P$ denotes projection. Observe first that since $\pi \circ \rho = \pi_P \circ \pi_1$,

$$\pi_* \tilde{\mathcal{H}} = (\pi \circ \rho)_*(\mathcal{H} \times \{0\}) = \pi_{P*} \circ \pi_{1*}(\mathcal{H} \times \{0\}) = \pi_{P*}\mathcal{H} = TM.$$

Thus, $T(P \times_G F) = \ker \pi_* \oplus \tilde{\mathcal{H}}$, and uniqueness is immediate, since $\ker \pi_* \perp \tilde{\mathcal{H}}$, and the inner product is specified on each factor. To establish existence, given $(p, m) \in P \times F$, endow $\tilde{\mathcal{H}}_{\rho(p,m)}$ with the inner product for which the restriction $\pi_* : \tilde{\mathcal{H}}_{\rho(p,m)} \to M_{\pi_P(p)}$ becomes a linear isometry. Next, endow the vertical space $\ker \pi_{*\rho(p,m)} = \rho_*(\{0\} \times F_m)$ with the inner product for which $h_{p*} : F_m \to \rho_*(\{0\} \times F_m)$ becomes a linear isometry, where $h_p : F \to \rho(p, F)$ is given by $h_p(m) = \rho(p, m)$; i.e., we endow the fiber $\rho(p, F)$ over $\pi_P(p)$ with the Riemannian metric for which h_p becomes an isometry. To see that this metric is well defined, observe that if $L_g : F \to F$ denotes the isometric action of $g \in G$ on F, then $h_{pg} = h_p \circ L_g$. Thus, h_p is an isometry iff h_{pg} is one for any $g \in G$, and the definition is independent of the point (p, m) chosen in the fiber. Finally, set $\ker \pi_* \perp \tilde{\mathcal{H}}$. By construction, $\pi : P \times_G F \to M$ is a Riemannian submersion, and it remains to show that the fibers are totally geodesic. Consider, to this end, a regular curve $c : [0, 1] \to M$ in M. If $\tilde{c}$ denotes its horizontal lift to P (meaning $\tilde{c}$ is tangent to $\mathcal{H}$) starting at p, then $\rho \circ (\tilde{c}, m)$ is the horizontal lift of c to $P \times_G F$ starting at $\rho(p, m)$. Now, $\mathcal{H}$ is a G-connection, so that if c is closed, then $\tilde{c}(1) = pg$ for some $g \in G$. Thus, $\rho(\tilde{c}(1), m) = \rho(pg, m) = \rho(p, gm)$, and under the identification $h_p : F \to \rho(p, F)$, the holonomy diffeomorphism induced by c is just the isometry L_g of F. But if holonomy transformations are isometries, then by the discussion in Section 1.4, the fibers are totally geodesic: In fact, a holonomy Jacobi field along a horizontal geodesic must have constant norm, so that

$$0 = |J|^{2\prime} = -2\langle S_{\dot{c}} J, J\rangle,$$

and $S \equiv 0$. $\square$

It is worth noticing that the construction outlined at the beginning of the section is actually a special case of Proposition 2.7.1, provided the fibers are totally geodesic; i.e., if G acts freely on the right by isometries on a Riemannian manifold P, acts isometrically on the left on F, and if $P \to P/G$ has totally geodesic fibers, then the metric on $P \times_G F$ for which $\rho : P \times F \to P \times_G F$ becomes a Riemannian submersion is a connection metric as in Proposition 2.7.1: Since the horizontal distribution $\mathcal{H}$ of the Riemannian submersion $\pi_P : P \to M = P/G$ is invariant under the action of G, it is a connection on the principal bundle π_G. By commutativity of the diagram

$$\begin{array}{ccc} P \times F & \xrightarrow{\rho} & P \times_G F \\ {\scriptstyle \pi_1}\downarrow & & \downarrow{\scriptstyle \pi} \\ P & \xrightarrow[\pi_P]{} & M \end{array},$$

π_* maps $\rho_*(\mathcal{H} \times \{0\})$ isometrically onto TM, so that $\rho_*(\mathcal{H} \times \{0\})$ is the horizontal distribution of π. Since the fibers are totally geodesic, the claim follows from the uniqueness part of the proposition. The fibers, though, will not, in general, be isometric to the original Riemannian manifold F. The point here is that connection metrics are equivalent to Riemannian submersions with totally geodesic fibers in the following sense:

Theorem 2.7.2. *Let $\pi : M \to B$ be a Riemannian submersion with totally geodesic fibers. Then π is a fiber bundle and the metric on M is a connection metric.*

Proof. That π is a fiber bundle was the content of Theorem 1.4.1 (1). In fact, recall that for fixed $b_0 \in B$, if $F := \pi^{-1}(b_0)$ and G is the Lie group of isometries of F, then the corresponding principal G-bundle $P \to B$ has as fiber $\pi_P^{-1}(b)$ the collection of all isometries $F \to \pi^{-1}(b)$, and M is identified with $P \times_G F$ via

$$\begin{aligned} P \times_G F &\longrightarrow M, \\ \rho(h, q) &\longmapsto h(q). \end{aligned}$$

By definition, given $h \in \pi_P^{-1}(b)$, the map

$$\begin{aligned} F &\longrightarrow \rho(h, F) = \pi^{-1}(b), \\ q &\longmapsto \rho(h, q) \end{aligned}$$

is just h, hence is an isometry. It remains to show that under the identification $M = P \times_G F$, the horizontal distribution $\tilde{\mathcal{H}}$ of $M \to B$ equals $\rho_*(\mathcal{H} \times \{0\})$ for some connection $\mathcal{H}$ on P; i.e., that $R_{g*}\mathcal{H}_h = \mathcal{H}_{hg}$ for $h \in P$, $g \in G$, or equivalently, that if $\gamma : I \to P$ is a curve in P such that $\rho(\gamma, q)$ is horizontal in M, then $\rho(R_g \circ \gamma, q)$ is also horizontal for $g \in G$. Now, if $c : [0, 1] \to B$ is a curve with $c(0) = b$, then the horizontal lift of c in M starting at some point $\rho(h, q) \in \pi^{-1}(b)$ is given by

$$t \longmapsto \rho(h_{c|_{[0,t]}} \circ h, q),$$

where h_c is the holonomy transformation associated to c. But then for any $g \in G$ the curve

$$t \longmapsto \rho(R_g \circ h_{c|_{[0,t]}} \circ h, q) = h_{c|_{[0,t]}}(h(gq))$$

is the horizontal lift of c starting at $g(q)$, thereby establishing the claim. $\square$

Example 2.7.2. Let G be a Lie group, K, H compact subgroups with $K \subset H$. Then there exist G-invariant metrics on G/K, G/H, and a normal homogeneous metric on H/K such that the natural fibration $\pi : G/K \to G/H$ becomes a Riemannian submersion with totally geodesic fibers isometric to H/K, cf. [15]. To see this, choose a left-invariant metric on G that is right-invariant under H, so that $\pi_G : G \to G/H$ is a Riemannian submersion as in Section 2.5. We claim that the horizontal distribution $\mathcal{H}$ of π_G is a connection on the principal H-bundle π_G: given $h \in H$, $g \in G$, we have $R_{h*}\mathcal{H}_g = \mathcal{H}_{gh}$, since R_h is an isometry of G that preserves the fibers, and therefore also their orthogonal complement. Thus, $\mathcal{H}$ is a connection (this also follows from the fact that the fibers of $G \longrightarrow G/H$ are totally geodesic – by Theorem 2.4.1 – together with Theorem 2.7.2). The restriction of the metric to H is bi-invariant, so that H acts by isometries on the normal homogeneous space H/K. By Proposition 2.7.1, there exists a metric on the total space $G/K = G \times_H H/K$ of the associated bundle with fiber H/K for which $\pi : G/K \to G/H$ becomes a Riemannian submersion with totally geodesic fibers. It is straightforward to verify that this metric is G-invariant: left translation $L_g : G \to G$ induces a well-defined diffeomorphism $\mathbb{L}_g$ of $G \times_H H/K$ such that the diagram

$$\begin{array}{ccc} G \times H/K & \xrightarrow{L_g \times 1_{H/K}} & G \times H/K \\ \rho \downarrow & & \downarrow \rho \\ G \times_H H/K & \xrightarrow[\mathbb{L}_g]{} & G \times_H H/K \end{array}$$

commutes. Furthermore, since L_g also induces an isometry of G/H, we have that $L_{g*}\mathcal{H} = \mathcal{H} \circ L_g$. Thus,

$$\mathbb{L}_{g*}\rho_*(\mathcal{H} \times \{0\}) = \rho_*(L_{g*}\mathcal{H} \times \{0\}) = \rho_*(\mathcal{H} \circ L_g \times \{0\}),$$

and $\mathbb{L}_g$ preserves the horizontal distribution of π. Since L_g is an isometry, the restriction of $\mathbb{L}_{g*}$ to this horizontal distribution is isometric. Finally, the restriction to the vertical distribution is also isometric, because if $a \in G$, $h \in H$, and $(0, u) \in G_a \times (H/K)_{hK}$, then

$$\mathbb{L}_{g*}\rho_{*(a,hK)}(0, u) = \rho_*(L_{g*}0, u) = \rho_{*(ga,hK)}(0, u),$$

and $|\rho_{*(a,hK)}(0, u)| = |\rho_{*(ga,hK)}(0, u)| = u$, by definition of the fiber metric.

Proposition 2.7.1 enables us to construct metrics of positive Ricci curvature on certain fiber bundles, following results of Poor [106], Nash [96], and Bérard

Bergery [17]. But first, some notation: given a Riemannian manifold M, denote as in Section 2.4 by ρ the Ricci form of the metric,

$$\rho(x) := \mathrm{Ric}(x,x), \qquad x \in TM.$$

If $\pi : M \longrightarrow B$ is a Riemannian submersion, and $u \in \mathcal{V}$, define

$$\begin{aligned} A_u : \mathcal{H} &\longrightarrow \mathcal{H}, \\ x &\longmapsto A_x^* u. \end{aligned}$$

The following is an immediate consequence of Theorem 1.5.1:

Lemma 2.7.1. *Let $\pi : M \to B$ be a Riemannian submersion with totally geodesic fibers. Then for basic X,*

$$\begin{aligned} (\rho_M - \rho^{\mathbf{h}})(X) &= 3 \operatorname{tr} A_X^* A_X; \\ \rho^{\mathbf{v}}(X) = (\rho - \rho^{\mathbf{h}})(X) &= \operatorname{tr} A_X A_X^* = \operatorname{tr} A_X^* A_X; \\ \rho^{\mathbf{h}}(T) &= \operatorname{tr} A_T^* A_T, \end{aligned}$$

where, as usual, X also denotes the π-related vector field on B.

Theorem 2.7.3. *Let M and F denote compact Riemannian manifolds with positive Ricci curvature, $\pi : E \to M$ a fiber bundle with fiber F and structure group G. If the metric on F is G-invariant, then E admits a metric with positive Ricci curvature.*

Proof. Endow E with some connection metric as in Proposition 2.7.1, and apply vertical warping to the fiber as in Section 2.1, taking the function ϕ to be a constant $r \in \mathbb{R}$. If $\tilde{\rho}$ denotes the Ricci curvature of the new metric, then decomposing a vector field $W = X + T$ on E as a sum of horizontal X and vertical T, we have

$$\tilde{\rho}(W) = \tilde{\rho}(X) + \tilde{\rho}(T) + 2\widetilde{\mathrm{Ric}}(X,T). \tag{2.7.2}$$

(2.1.28) yields

$$\tilde{\rho}(X) = \rho(X) + \frac{2}{3}(1 - e^{2r})(\rho_M - \rho^{\mathbf{h}})(X),$$

which, by Lemma 2.7.1, becomes

$$\begin{aligned} \tilde{\rho}(X) &= e^{2r}\rho(X) + (1 - e^{2r})\rho(X) + (1 - e^{2r})\rho_M(X) - \frac{1}{3}(1 - e^{2r})\rho_M(X) \\ &\quad - \frac{2}{3}(1 - e^{2r})\rho^{\mathbf{h}}(X) \\ &= (1 - e^{2r})\rho_M(X) + e^{2r}\rho(X) + (1 - e^{2r})(\rho - \rho^{\mathbf{h}})(X) \\ &\quad - \frac{1}{3}(1 - e^{2r})(\rho_M - \rho^{\mathbf{h}})(X) \\ &= (1 - e^{2r})\rho_M(X) + e^{2r}\rho(X). \end{aligned} \tag{2.7.3}$$

Similarly, (2.1.34) yields

$$\tilde{\rho}(T) = (1 - e^{2r})\{\rho_F(T) - e^{2r} \operatorname{tr} A_T^* A_T\} + e^{2r} \rho(T),$$

which, by the lemma, becomes

$$\tilde{\rho}(T) = \rho_F(T) + e^{4r} \rho^{\mathbf{h}}(T). \tag{2.7.4}$$

Finally, the Codazzi equation implies that $\operatorname{Ric}^{\mathbf{v}}(X,T) = 0$ because the fibers are totally geodesic, and (2.1.31) becomes

$$\widetilde{\operatorname{Ric}}(X,T) = e^{2r} \operatorname{Ric}^{\mathbf{h}}(X,T). \tag{2.7.5}$$

Substitution of (2.7.3), (2.7.4), and (2.7.5) in (2.7.2) then yields

$$\tilde{\rho}(W) = (1 - e^{2r})\rho_M(X) + e^{2r}\rho(X) + \rho_F(T) + e^{4r}\rho^{\mathbf{h}}(T) + 2e^{2r} \operatorname{Ric}^{\mathbf{h}}(X,T).$$

As $r \to -\infty$, $\rho(W) \to \rho_M(X) + \rho_F(T) > 0$, and the result follows by compactness of E. □

Notice that the theorem applies to the principal bundle $\pi_P : P \longrightarrow M$ of any vector bundle of rank $k \geq 3$ over a compact manifold M of positive Ricci curvature, since for $k > 2$, the orthogonal group $O(k)$ admits a bi-invariant metric of positive Ricci curvature. It does not apply, however, to the one-dimensional $O(2)$. In this case, we argue as in [17]: Up to a 2-fold cover and scaling, we may assume that π_P is a principal S^1-bundle. Let T be a basis of the Lie algebra of S^1, and $\tilde{T}$ the corresponding fundamental vector field, cf. (2.3.3). The curvature form Ω of a connection on π_P may then be identified with the real-valued 2-form (also denoted Ω) on P given by

$$\Omega(p)(X,Y) := \langle \Omega(p)(X,Y), T\rangle, \qquad X, Y \in \mathcal{H}, \quad p \in P,$$

and is always the pullback via π_P of a closed 2-form on M, the cohomology class of which is $2\pi e \in H^2(M,\mathbb{R})$, where e is the real Euler class of the bundle. Denote by α the unique harmonic representative in that cohomology class. By [82], there exists a connection on π_P whose curvature form equals $\pi_P^*\alpha$. Endow P with the corresponding connection metric, with totally geodesic fibers of length 2π. (2.3.2) then implies that

$$\alpha(X,Y) = -2\langle A_{\bar{X}}\bar{Y}, \tilde{T}\rangle, \tag{2.7.6}$$

where $\bar{X}$, $\bar{Y}$ are the basic lifts of X, $Y \in \mathfrak{X}M$. Observe that if ϕ_t denotes the flow of the vertical Killing field $\tilde{T}$, then any basic field $\bar{X}$ is ϕ_t-related to itself, so that $[\bar{X}, \tilde{T}] = 0$. Thus,

$$\nabla^{\mathbf{v}}_{\bar{X}} \tilde{T} = \nabla^{\mathbf{v}}_{\tilde{T}} \bar{X} = -S_{\bar{X}} \tilde{T} = 0.$$

If X_i is a local orthonormal basis of $\mathfrak{X}M$, harmonicity of α implies that for $X \in \mathfrak{X}M$,

$$\begin{aligned}
0 = \delta\alpha(X) &= \sum_i (\nabla_{X_i}\alpha)(X_i, X) \\
&= \sum_i \{\nabla_{X_i}(\alpha(X_i, X)) - \alpha(\nabla_{X_i}X_i, X) - \alpha(X_i, \nabla_{X_i}X)\} \\
&= -2\sum_i \{\bar{X}_i\langle A_{\bar{X}_i}\bar{X}, \tilde{T}\rangle - \langle A_{\nabla_{\bar{X}_i}\bar{X}_i}\bar{X}, \tilde{T}\rangle - \langle A_{\bar{X}_i}\nabla_{\bar{X}_i}\bar{X}, \tilde{T}\rangle\} \\
&= -2\sum_i \{\langle\nabla_{\bar{X}_i}(A_{\bar{X}_i}\bar{X}), \tilde{T}\rangle - \langle A_{\nabla_{\bar{X}_i}\bar{X}_i}\bar{X}, \tilde{T}\rangle - \langle A_{\bar{X}_i}\nabla_{\bar{X}_i}\bar{X}, \tilde{T}\rangle\} \\
&= -2\sum_i \langle(\nabla_{\bar{X}_i}A)_{\bar{X}_i}\bar{X}, \tilde{T}\rangle = -2\sum_i \langle R(\bar{X}_i, \bar{X})\bar{X}_i, \tilde{T}\rangle,
\end{aligned}$$

where the last equality uses (1.5.7). It follows that $\mathrm{Ric}^{\mathbf{h}}(\tilde{X}, \tilde{T}) \equiv 0$, and the formula for the Ricci form $\tilde{\rho}(W)$ at $W = X + T$ in the warped metric becomes

$$\tilde{\rho}(W) = (1 - e^{2r})\rho_M(X) + e^{2r}\rho(X) + e^{4r}\rho^{\mathbf{h}}(T).$$

It follows that the Ricci curvature can be made nonnegative, and strictly positive at any point where

$$\rho^{\mathbf{h}}(T) = \operatorname{tr} A_T^* A_T \geq 0$$

is nonzero for $T \neq 0$. This implies that if the Euler class is nonzero, then $\alpha \neq 0$, and by (2.7.6), there is one point at least where the Ricci curvature will be positive. But by a result of Aubin [4], such a metric can be deformed to one of strictly positive Ricci curvature. On the other hand, if the Euler class is zero, then the bundle is trivial, and P is diffeomorphic to $M \times S^1$. By Myers' theorem, P cannot admit a metric of positive Ricci curvature. Summarizing, we have:

Theorem 2.7.4. *Let $\pi_P : P \to M$ be a principal $O(k)$-bundle over a compact manifold M of positive Ricci curvature. If $k > 2$, then P admits a metric of positive Ricci curvature. When $k = 2$, P admits a metric of positive Ricci curvature iff π_P or its orientable 2-fold cover is not a trivial bundle.*

Remark 2.7.2. Nash [96] has shown that if the total space of a principal $O(k)$-bundle over a compact M with positive Ricci curvature admits a metric with positive Ricci curvature, then so does the total space of the associated rank k vector bundle over M. Thus, *the total space of any vector bundle of rank $k \geq 2$ over a compact manifold of positive Ricci curvature admits a complete metric of positive Ricci curvature*: This follows from Theorem 2.7.4 unless the bundle, up to a 2-fold cover, is trivial of rank two. But if P denotes $\mathbb{R}^2$ with a metric of positive sectional curvature, then $M \times P$ with the product metric has positive Ricci curvature. On the other hand, the statement is not true for rank one bundles: up to a 2-fold cover, the total space of such a bundle is diffeomorphic to $M \times \mathbb{R}$, and as such cannot admit a metric of positive Ricci curvature. Indeed, for any complete metric on $M \times \mathbb{R}$, there must exist a line; i.e., a normal geodesic $c : \mathbb{R} \to M$ with $d(c(t), c(t')) = |t - t'|$ for all t, t'; one way to see this is to fix some $p \in M$ and observe that for any $n \in \mathbb{N}$ there exists a minimal normal geodesic c_n from $(p, -n)$ to(p, n). c_n intersects $M \times \{0\}$ in a unique point $c_n(t_n)$. Since the sequence $v_n := \dot{c}_n(t_n)$ lives in a compact set (namely, the restriction of the unit tangent bundle of $M \times \mathbb{R}$ to $M \times \{0\}$), we may assume it converges to some v. It is now easy to see that the geodesic $t \mapsto \exp(tv)$ is a line. Gromoll and Meyer, however, have shown that a manifold of positive Ricci curvature cannot contain a line [58], cf. also [36].

2.8 Fat bundles

Given a submersion $\pi : M \to B$, one usually seeks to endow B with a metric of positive (or nonnegative) curvature by projecting via π an existing metric on M. But the problem can also be inverted: If B is a Riemannian manifold with positive curvature, one may ask whether there is a metric of positive curvature on M for which π becomes Riemannian. Now, the curvature of a *vertizontal* plane (i.e., a plane spanned by a unit vertical u and unit horizontal x) is

$$K_{x,u} = \langle (\nabla^{\mathrm{v}}_x S)_x u, u\rangle - |S_x u|^2 + |A_x^* u|^2$$

by Theorem 1.5.1. The easiest way to guarantee that this be nonnegative is to require that the fibers be totally geodesic. This motivates the following definition:

Definition 2.8.1. A Riemannian submersion with totally geodesic fibers is said to be *fat* if all vertizontal planes have positive curvature.

This terminology was introduced by Weinstein [137], see also [146] for a comprehensive survey of fat bundles. The reason the words 'bundle' and 'submersion' are often used interchangeably when referring to fatness is that according to Theorem 2.7.2, any Riemannian submersion with totally geodesic fibers is a fiber bundle with a connection metric. Furthermore, we will soon see that fatness can be expressed in terms of the connection.

By definition, a submersion is fat if $A_X^* : \mathcal{V} \to \mathcal{H}$ is 1-1, or equivalently, if $A_X : \mathcal{H} \to \mathcal{V}$ is onto for any nonzero $X \in \mathcal{H}$. This already imposes fairly stringent restrictions on the dimensions of the spaces involved:

Lemma 2.8.1. *Let $\pi : M \to B$ denote a fat submersion. Then*

1. *B is even-dimensional;*
2. $\dim \mathcal{V} \leq \dim \mathcal{H} - 1$, *and equality only occurs when* $\dim \mathcal{H} = 2, 4,$ *or* 8.

Proof. At any $p \in M$, for fixed nonzero $u \in \mathcal{V}_p$, the assignment $(x, y) \mapsto \langle A_x y, u\rangle$ is a skew-symmetric nondegenerate bilinear form on $\mathcal{H}_p$; i.e., a symplectic form. Elementary linear algebra implies that $\mathcal{H}$, and hence also B, is even-dimensional.

The second statement follows by observing that the assignment $x \mapsto A_x^* u$ defines a nowhere zero vector field on the unit sphere in $\mathcal{H}_p$. Fatness actually yields dim $\mathcal{V}$ linearly independent vector fields on the sphere of dimension dim $\mathcal{H}-1$. This clearly implies the inequality. Equality can only occur if the sphere is parallelizable, in which case it must have dimension one, three, or seven. □

By Theorem 2.7.2, if $\pi : M \to B$ is fat, then π is an associated G-bundle, with G denoting the isometry group of the fiber, and $M = P \times_G F$. Since the $A_X Y$ are vertical Killing fields that span the fiber at each point, the action of G on a fiber F must be transitive, so that F is a homogeneous space G/H, where H denotes the isotropy group at some point of F. Thus, M itself is diffeomorphic to $P \times_G (G/H) = P/H$.

Our next objective is to examine the restrictions that fatness imposes on the connection, beginning with a connection $\mathcal{H}$ on the principal bundle. If ω and Ω denote the connection and curvature forms of $\mathcal{H}$, then by definition,

$$\omega(A_X Y) = -\frac{1}{2}\Omega(X, Y)$$

for horizontal fields X and Y on P, cf. (2.3.2). We assume from now on that G is endowed with some fixed bi-invariant metric, or alternatively, that $\mathfrak{g}$ is endowed with an Ad-invariant inner product. Then fatness of the connection is equivalent to non-degeneracy of the 2-form

$$(x, y) \longmapsto \langle \Omega(x, y), u \rangle$$

on $\mathcal{H}$ for every $u \in \mathfrak{g}$. More generally, any u for which the above form is nondegenerate is called a *fat vector*. Notice that if u is fat, then so is the orbit of u under the adjoint action of G: this follows from the identity

$$R_g^* \Omega = \operatorname{Ad}_{g^{-1}} \Omega, \qquad g \in G,$$

which together with bi-invariance of the metric on G implies

$$\langle \Omega(X, Y), \operatorname{Ad}_g u \rangle = \langle \operatorname{Ad}_{g^{-1}} \Omega(X, Y), u \rangle = \langle \Omega(R_{g*}X, R_{g*}Y), u \rangle.$$

Recall that the integral of a function $f : G \to \mathbb{R}$ on an oriented Lie group G is defined to be

$$\int_G f := \int_G f\mu,$$

where μ is the (dim G)-bi-invariant form on G consistent with the orientation that satisfies $\int_G \mu = 1$. It is a standard fact that for left translation L_g by g,

$$\int_G f = \int_G f \circ L_g,$$

cf. [136]. Fix a fat vector u, and define a polynomial p on $\mathfrak{g}$ by

$$p(w) = \int_G \langle w, \operatorname{Ad}_g u\rangle^n,$$

where $n = (\dim B)/2$. In other words $p(w)$ is the integral of the function $\phi : G \to \mathbb{R}$, where $\phi(g) = \langle w, \operatorname{Ad}_g u\rangle^n$. Then p is Ad-invariant, because

$$\begin{aligned} p(\operatorname{Ad}_h w) &= \int_G \langle \operatorname{Ad}_h w, \operatorname{Ad}_g u\rangle^n = \int_G \langle w, \operatorname{Ad}_{h^{-1}g} u\rangle^n = \int_G \phi \circ L_{h^{-1}} = \int_G \phi \\ &= p(w). \end{aligned}$$

By Chern-Weil theory, the $2n$-form $p(\Omega)$ on P, with

$$\begin{aligned} p(\Omega)(X_1, \dots, X_{2n}) = \frac{1}{(2n)!} \sum_\sigma (\operatorname{sgn} \sigma) \int_G \langle \Omega(X_{\sigma(1)}, X_{\sigma(2)}), \operatorname{Ad}_g u\rangle \cdots \\ \cdots \langle \Omega(X_{\sigma(2n-1)}, X_{\sigma(2n)}), \operatorname{Ad}_g u\rangle \end{aligned}$$

is the pull-back to P of a closed $(2n)$-form α on B, the cohomology class of which is independent of the choice of connection, and represents a characteristic class of the bundle.

Theorem 2.8.1 (Weinstein). *Let $u \in \mathfrak{g}$ be a fat vector for some connection on $\pi_P : P \to B$. Then the bundle has at least one nonzero characteristic number.*

Proof. For $g \in G$, denote by μ_g the one-form on $\mathfrak{g}$ given by $\mu_g(v) = \langle v, \operatorname{Ad}_g u\rangle$, $v \in \mathfrak{g}$. Then

$$p(\Omega) = \int_G (\mu_g \circ \Omega) \wedge \cdots \wedge (\mu_g \circ \Omega),$$

and by fatness, $\mu_g \circ \Omega$ is the pull-back of a symplectic (i.e., nondegenerate) form on B. It follows that $(\mu_g \circ \Omega)^n$, and hence also $p(\Omega)$ is the pull-back of a volume form on B; in other words, the characteristic class above is represented by a volume form α, so that the corresponding characteristic number $\int_B \alpha \neq 0$. $\square$

Example 2.8.1. If $G = S^1$, then $\Omega = 2\pi\pi_P^* e$, where e represents the Euler class of π_P. Since $[\Omega^n] \neq 0$, $[e^n] \neq 0$. Similarly, it can be shown that if $G = S^3$ or $SO(3)$, then $\dim B = 4n$, and $[\alpha^n] \neq 0$, where $[\alpha]$ denotes the first Pontrjagin class of the bundle, see [146].

We next look at fatness in the context of associated bundles, which we have seen are of the form $\pi : M = P \times_G (G/H) \to B$, where H is the isotropy group of some $m_0 \in G/H$. For $p \in P$, $m \in G/H$, denote by $\imath_p : G/H \to P \times G/H$ and $\imath_m : P \to P \times G/H$ the maps $q \mapsto (p, q)$ and $r \mapsto (r, m)$, respectively, and by l_p (resp. r_m) the action of G on $p \in P$ (resp. on $m \in G/H$). If $\rho : P \times (G/H) \to M$ is projection, then $\rho \circ \imath_m \circ l_p = \rho \circ \imath_p \circ r_m$. Furthermore, the vertical space at p of the principal bundle $\pi_P : P \to B$ is $l_{p*e}\mathfrak{g}$. Thus, if u is a vertical vector in T_pP, then

$$\begin{aligned}\rho_*(u,0) &= \rho_*(0, r_{m*}l_{p*e}^{-1}u) = \rho_* \circ \imath_{m*}u = \rho_* \circ \imath_{p*} \circ r_{m*} \circ l_{p*e}^{-1}u \\ &= \rho_*(0, r_{m*}\omega(u)).\end{aligned}$$

The above formula enables us to compute the A-tensor: Let X, $Y \in \mathfrak{X}B$, with basic lifts $\tilde{X}$, $\tilde{Y}$ to P. The corresponding basic lifts to M are $\bar{X} = \rho_*(\tilde{X},0)$ and $\bar{Y} = \rho_*(\tilde{Y},0)$. Thus,

$$\begin{aligned}A_{\bar{X}}\bar{Y} &= \frac{1}{2}[\bar{X},\bar{Y}]^{\mathrm{v}} = \frac{1}{2}(\rho_*[\tilde{X},\tilde{Y}]^{\mathrm{v}},0) = \frac{1}{2}\rho_*(0, r_{m*}\omega[\tilde{X},\tilde{Y}]) \\ &= -\frac{1}{2}\rho_*(0, r_{m*}\Omega(\tilde{X},\tilde{Y})).\end{aligned}$$

Now, the restriction $r_{m*e} : \mathfrak{h}^{\perp} \to (G/H)_m$ is isometric, and by definition of the connection metric, the map

$$\begin{aligned}\rho_p : G/H &\longrightarrow \rho(p, G/H), \\ gH &\longmapsto \rho(p, gH)\end{aligned}$$

is an isometry. It follows that the associated bundle π is fat if and only if

$$\langle \Omega(\tilde{X},\tilde{Y}), u\rangle = -2\langle A_{\bar{X}}\bar{Y}, \rho_{p*}r_{m*}u\rangle \neq 0$$

for every $u \in \mathfrak{h}^{\perp}$, a situation we formally characterize below:

Definition 2.8.2. Let H be a subgroup of a compact group G. A connection on a principal G-bundle is said to be H-*fat* if

$$(X,Y) \longmapsto \langle \Omega(X,Y), u\rangle$$

is nondegenerate for all nonzero $u \in \mathfrak{h}^{\perp}$.

We have proved:

Proposition 2.8.1. *A connection on an associated bundle $P \times_G (G/H) \to B$ is fat iff the corresponding connection on the principal G-bundle $P \to B$ is H-fat.*

We wish to illustrate Proposition 2.8.1 in the important special case of a sphere bundle. Before doing so, let us recall without proof some facts concerning the identification of the orthogonal algebra with the space of bivectors. For further details, see for example [136].

If E is an inner product space, denote by $\mathfrak{o}(E)$ the Lie algebra of skew-adjoint transformations of E. There is a canonical isomorphism $\mathcal{I} : \Lambda_2(E) \to \mathfrak{o}(E)$, which on decomposable elements is given by

$$\mathcal{I}(u \wedge v)(w) = \langle v, w\rangle u - \langle u, w\rangle v.$$

Notice that when u and v are orthonormal, this is a rotation by $\pi/2$ in the plane spanned by u and v. Endow $\mathfrak{o}(E)$ with the inner product

$$\langle A, B\rangle = -1/2 \operatorname{tr} AB,$$

and $\Lambda_2(E)$ with the one for which $\mathcal{I}$ becomes a linear isometry. It follows that if e_i is an orthonormal basis of E, then $e_i \wedge e_j$, $i < j$, is an orthonormal basis of $\Lambda_2(E)$.

The Lie group $O(E)$ of orthogonal transformations of E acts on $\mathfrak{o}(E)$ via the adjoint action. Under the isomorphism $\mathcal{I}$, the action of $O(E)$ on $\Lambda_2(E)$ is given by

$$A(u \wedge v) = Au \wedge Av, \qquad A \in O(E), \quad u, v \in E, \tag{2.8.1}$$

and extending linearly. Moreover, for $M \in \mathfrak{o}(E)$,

$$\langle M, u \wedge v\rangle = \langle Mv, u\rangle, \qquad u, v \in E. \tag{2.8.2}$$

With these preliminaries out of the way, consider a sphere bundle $S^k \to M \xrightarrow{\pi} B$, with corresponding principal $O(k+1)$-bundle $P \to B$. There is an associated vector bundle $\mathbb{R}^{k+1} \to E = P \times_{O(k+1)} \mathbb{R}^{k+1} \to B$, which inherits a connection from one on the principal bundle. The following result expresses fatness of the sphere bundle in terms of the curvature tensor R of the connection on the vector bundle:

Proposition 2.8.2. *A sphere bundle is fat iff the 2-form*

$$\begin{aligned} B_m \times B_m &\longrightarrow \mathbb{R}, \\ (x, y) &\longmapsto \langle R(x, y)u, v\rangle \end{aligned}$$

is nondegenerate for all linearly independent u, $v \in E_m$, $m \in B$.

Proof. The fiber over m of the total space P of the principal $O(k+1)$-bundle $\pi : P \to B$ consists of all orthonormal bases of E_m. If b denotes such a basis, then for x, y in the tangent space of P at b, the matrix of $R(\pi_*x, \pi_*y) \in \mathfrak{o}(E_m)$ with respect to the basis b is $\Omega(b)(x, y)$, cf. [136]. Since the total space of the sphere bundle is $M = P \times_{O(k+1)} O(k+1)/O(k)$, Proposition 2.8.1 says that $M \to B$ is fat iff the 2-form

$$(x, y) \longmapsto \langle \Omega(b)(x, y), \alpha\rangle = \langle R(\pi_*x, \pi_*y), b(\alpha)\rangle$$

is nondegenerate for all nonzero $\alpha \in \mathfrak{o}(k)^{\perp}$, $b \in P$. Here, the basis b is viewed as a linear isometry $b : \mathbb{R}^{k+1} \to E_m$ that extends in a natural way to an isometry $\mathfrak{o}(k+1) \to \mathfrak{o}(E_m)$ via the identification $\Lambda_2(\mathbb{R}^{k+1}) \cong \mathfrak{o}(k+1)$; i.e., $b(u \wedge v) = bu \wedge bv$ for u, $v \in \mathbb{R}^{k+1}$. If we identify $\mathfrak{o}(k)$ with the subspace of $\mathfrak{o}(k+1)$ spanned by $e_i \wedge e_j$ for $1 < i < j$, then $e_1 \wedge e_2 \in \mathfrak{o}(k)^{\perp}$. Now, recall that if $e_1 \wedge e_2$ is fat, then so is $\mathrm{Ad}_g(e_1 \wedge e_2)$ for any $g \in O(k+1)$. By (2.8.1), this means that the 2-form above is nondegenerate for any $\alpha = \mathrm{Ad}_g\, e_1 \wedge \mathrm{Ad}_g\, e_2$, $g \in O(k+1)$; i.e., for any decomposable vector in $\mathbb{R}^{k+1}$. In terms of R, this amounts to nondegeneracy of

$$(x, y) \mapsto \langle R(\pi_* x, \pi_* y), \beta \rangle$$

for any decomposable $\beta \in \Lambda_2(E_m)$. Together with (2.8.2), this establishes the result. □

The simplest examples of fat sphere bundles are of course the Hopf fibrations, since they have totally geodesic fibers and their total spaces are positively curved. In fact, it was shown in [40] that among all S^3-bundles over S^4, the Hopf fibration is the only fat bundle. Another indication that fat bundles are scarce is the fact that, when the fiber dimension is larger than 1, the only known examples are bundles of the form $H/K \to G/K \to G/H$ as in Example 2.7.2 (a fat S^1-bundle is just one with symplectic base). All fat bundles of that type were classified in [15]. It is actually easy to give a criterion for fatness in this case:

Proposition 2.8.3. *Let G be a Lie group with bi-invariant metric. Then the bundle $\pi : G/K \to G/H$ from Example 2.7.2 is fat iff $[X, U] \neq 0$ for all $0 \neq X \in \mathfrak{h}^\perp$ and $0 \neq U \in \mathfrak{k}^\perp \cap \mathfrak{h}$.*

Proof. By Proposition 2.8.1, π is fat iff the principal H-bundle $G \to G/H$ is K-fat; i.e., iff the 2-form

$$\begin{aligned} \mathfrak{h}^\perp \times \mathfrak{h}^\perp &\longrightarrow \mathbb{R}, \\ (X, Y) &\longmapsto \langle \Omega(X, Y), U \rangle \end{aligned}$$

is nondegenerate for all $U \in \mathfrak{k}^\perp \cap \mathfrak{h}$. But $\Omega(X, Y) = -[X, Y]_{\mathfrak{h}}$ by definition of the connection on $G \to G/H$, and

$$\langle \Omega(X, Y), U \rangle = -\langle [X, Y], U \rangle = \langle [X, U], Y \rangle$$

since the metric is bi-invariant. The claim clearly follows. □

Chapter 3

Open Manifolds of Nonnegative Curvature

Noncompact manifolds with a complete metric of nonnegative sectional curvature were studied in detail by Gromoll-Meyer [58], and by Cheeger-Gromoll [36], who gave a thorough account of their topology. Apart from some special cases, however, their metric structure has only been understood fairly recently. It illustrates the key role that Riemannian submersions seem to play in nonnegative curvature.

3.1 Convex sets in Riemannian manifolds

We begin by discussing several types of convexity that can occur in a connected Riemannian manifold M with distance function d. As usual, $B_\epsilon(p) = \{q \in M \mid d(p,q) < \epsilon\}$ will denote the metric ball of radius $\epsilon > 0$ centered at $p \in M$.

Definition 3.1.1. A subset C of M is said to be *convex* if any two points of C can be joined by a minimal geodesic of M, the image of which is contained in C. If, in addition, this geodesic is always unique in M, then C is said to be *strongly convex*.

A classical result of J.H.C. Whitehead states that for any $p \in M$ there exists a number $r(p) > 0$, called the *convexity radius at* p, such that if $\epsilon < r(p)$, then any metric ball contained in $B_\epsilon(p)$ is strongly convex. The corresponding boundary sphere has positive definite second fundamental form.

Definition 3.1.2. $C \subset M$ is said to be *locally convex* if for any p in the closure $\bar{C}$ of C, there exists $\epsilon(p) \in (0, r(p))$ such that $B_{\epsilon(p)}(p) \cap C$ is strongly convex.

Thus, a convex set is always locally convex, and a strongly convex set is convex. There is one further type of convexity that usually occurs in conjunction with certain functions: A function $f : M \to \mathbb{R}$ is said to be *convex* if for any geodesic $c : I \to M$, the function $f \circ c : I \to \mathbb{R}$ is convex in the usual sense. If f is such a function, consider a sublevel set $C^a = f^{-1}(-\infty, a]$, $a \in \mathbb{R}$, and points p, $q \in C^a$. Given any (not necessarily minimal) geodesic $c : [0,1] \to M$ from p to q, we have

$$(f \circ c)(t) \leq \max\{(f \circ c)(0), (f \circ c)(1)\}$$

by convexity of f, so that the image of c is contained in C^a. A subset C of M is said to be *totally convex* if any geodesic of M joining two points of C lies entirely inside C. Such a set is necessarily convex, provided M is complete. Proper totally convex subsets exist only in special situations: it is easy to see that a round sphere, for example, admits none. In fact, a result of Bangert implies that if M admits such a set, then it is noncompact [11].

The next theorem shows that even the weakest notion of convexity in an arbitrary Riemannian manifold shares features similar to those in Euclidean space:

Theorem 3.1.1 (Cheeger-Gromoll). *Let C be a closed, connected, locally convex subset of a Riemannian manifold M^n. Then C is an imbedded k-dimensional submanifold of M with totally geodesic connected (relative) interior, and (possibly non-smooth and/or empty) boundary, where $0 \leq k \leq n$.*

Proof. For now, we do not assume that C is closed. Let $k \in \{0, \dots, n\}$ denote the largest integer such that the collection of all smoothly imbedded k-dimensional submanifolds of M contained in C is nonempty, and denote by N the union of this collection. We claim that N is a smooth, totally geodesic submanifold of M. To establish the first part of the claim, it suffices to show that for any given $p \in N$, there exists a neighborhood U of p in N, and a neighborhood V of p in M such that $N \cap V = U$. Now, by assumption, p belongs to some k-dimensional submanifold

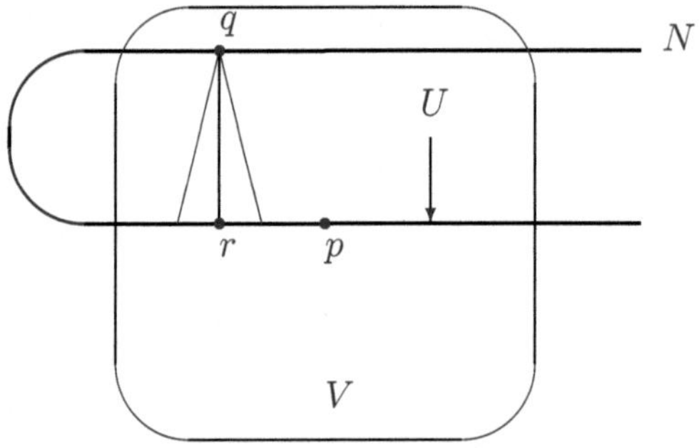

$N(p)$ of M contained in C. Consider a neighborhood $U \subset N(p) \cap B_{\epsilon(p)/2}(p)$ of p in $N(p)$, and choose $\delta \in (0, \epsilon(p)/2)$ small enough so that the exponential map of the normal bundle of U, when restricted to vectors of length less than δ, is a diffeomorphism onto a tubular neighborhood V of U in M. Then $N \cap V$ contains U, and hence equals U: for if $q \in (N \cap V) \setminus U$, and $r \in U$ is the point in U closest to q, then the minimal geodesic from q to r intersects U orthogonally. It follows that the minimal geodesic from q to any point in a sufficiently small neighborhood U_0 of r in U intersects U transversally. Then the cone

$$\{\exp(tu) \mid u \in M_q, |u| < \delta, 0 < t < 1, \exp u \in U_0\}$$

is a smooth $(k+1)$-dimensional submanifold of M which is contained in C by convexity. This contradicts the definition of k. Thus, N is a submanifold of M, and essentially the same argument shows it is totally geodesic.

Before proceeding any further with the proof, we point out the following:

Lemma 3.1.1. *Let $p \in C \cap \bar{N}$, q, $r \in B_{\epsilon(p)/4}(p)$, $\delta := d(q,r)$, and c a minimal normal geodesic from q to r. If $q \in N$ and $r \in C$, then $c[0,\delta) \subset N$, and in particular, $r \in \bar{N}$. Moreover, if $r \notin N$, then $c(\delta + t) \notin C$ for $0 < t < \epsilon(p)/4$.*

Proof. Since N is k-dimensional, there exists a $(k-1)$-dimensional hypersurface U in $N \cap B_{\epsilon(p)/4}(p)$ containing q and transversal to the image of c. Consider any $p_0 = c(\epsilon_0) \in C$ with $0 < \epsilon_0 < \delta + \epsilon(p)/4$. Then the smooth k-dimensional cone

$$V = \{\exp(tu) \mid u \in M_{p_0}, |u| < \epsilon(p), \exp(u) \in U, 0 < t < 1\}$$

must be contained in C, and hence also in N. In particular, choosing $p_0 = r$ implies that $c[0,\delta) \subset N$. Furthermore, if $c(t+\delta) \in C$, then choosing $p_0 = c(t+\delta)$ implies that $r = c(\delta) \in N$, which proves the last statement in the lemma. □

Resuming the proof of the theorem, we next claim that $C \subset \bar{N}$. To see this, consider a connected component N_0 of N. Then $C \subset \bar{N}_0$, for otherwise we can find points $p \in C \cap \bar{N}$, $q \in B_{\epsilon(p)/4}(p) \cap N_0$, and $r \in B_{\epsilon(p)/4}(p) \cap (C \setminus \bar{N}_0)$. But then, $r \in \bar{N}_0$ by Lemma 3.1.1, which is a contradiction. Since any connected component of N is dense in C, there can only be one; i.e., $N = N_0$, and $C \subset \bar{N}$ as claimed. Assume now that C is closed for the remainder of the argument, so that $C = \bar{N}$. Consider points $p \in \bar{N} \setminus N$, and $q \in B_{\epsilon(p)/4}(p) \cap N$. Define W to be the collection of all unit vectors u in N_q such that $\exp(su) \in (\bar{N} \setminus N) \cap B_{\epsilon(p)/4}(p)$ for some $s \in (0, \epsilon(p)/4)$. For any given $u \in W$, the value of s is unique by Lemma 3.1.1, and we denote it by $f(u)$. Lemma 3.1.1 also implies that W is open in the unit sphere in N_q, and that f is continuous on W. Then

$$\begin{aligned} F : (0,1] \times W &\to C, \\ (t,u) &\mapsto \exp_q t f(u) u \end{aligned}$$

is a homeomorphism onto a neighborhood of p in C. This completes the proof of the theorem. □

Since the boundary of a k-dimensional convex set is not necessarily smooth, there is in general no notion of a $(k-1)$-dimensional tangent space for a point in the boundary. We consider instead the following:

Definition 3.1.3. Let C be a closed convex set in M. The *tangent cone* at a point p in the boundary ∂C of C is the set

$$C_p = \{u \in M_p \mid \exp_p(tu) \in N \text{ for sufficiently small } t \in (0, \epsilon(p))\} \cup \{0\}.$$

Given any $q \in N$ at distance less than $\epsilon(p)/4$ from $p \in \partial C$, consider the minimal geodesic c from p to q. Then $\dot{c}(0)$ belongs to C_p, but by Lemma 3.1.1, $-\dot{c}(0)$ does not. Thus, $C_p \neq M_p$. However, given $u \in C_p \setminus \{0\}$, and $q := \exp(t_0 u) \in B_{\epsilon(p)/4}(p) \cap N$, the argument of Lemma 3.1.1 with the hypersurface of N through q orthogonal to c shows that there is an open neighborhood of u contained in C_p. Thus, $C_p \setminus \{0\}$ is open in M_p. This enables us to describe C_p in at least one important special case:

Proposition 3.1.1. *Let $C \subset M$ be closed, connected, and $p \in \partial C$. If there exists $q \in N$ and a minimal normal geodesic c from p to q that realizes the distance between q and the boundary of C, then $C_p \setminus \{0\}$ is the open half-space*

$$H = \{u \in M_p \mid \measuredangle(u, \dot{c}(0)) < \frac{\pi}{2}\}.$$

Proof. For $t_0 \in (0, \epsilon(p)/2)$ smaller than $d(p, q)$, $-c_{|[0,t_0]}$ also realizes the distance between $c(t_0)$ and the boundary of C, so that the closed ball of radius t_0 centered at $c(t_0)$ intersects ∂C in p only. A first variation of arc length argument then implies that $H \subset C_p$. On the other hand, if $v \in C_p \setminus \{0\}$, then $\measuredangle(v, \dot{c}(0)) \leq \pi/2$: otherwise, $-v \in H \subset C_p$, so that for sufficiently small t, both $\exp(tv)$ and $\exp(-tv)$ belong to N, and p itself then belongs to N by Lemma 3.1.1. Thus, $C_p \subset \bar{H}$. Since C_p is open in M_p, this completes the argument. □

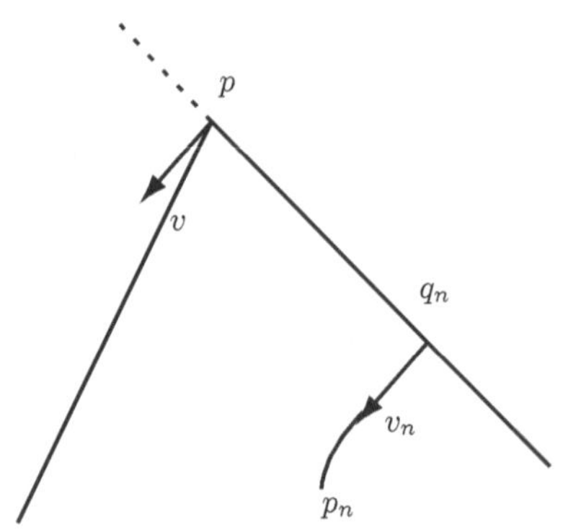

It can be shown that in general, $C_p \setminus \{0\}$ is an intersection of half-spaces. Notice that there always exists a half-space containing $C_p \setminus \{0\}$: In fact, let $p_n \to p$, $p_n \in N$. If q_n denotes the point of ∂C closest to p_n, then $q_n \to p$. Denote by v_n the initial tangent vector of the minimal normal geodesic from q_n to p_n. By Proposition 3.1.1, $C_{q_n} \setminus \{0\}$ is the open half-space $\{u \in M_{q_n} \mid \measuredangle(u, v_n) < \pi/2\}$ determined by v_n. $\{v_n\}$ may be assumed to converge to some unit vector $v \in M_p$, and it follows that the open half-space determined by v contains $C_p \setminus \{0\}$.

3.2 The soul construction

So far, we have made no assumption on the curvature of M. From now on, M will denote a complete, noncompact (open for short) manifold with sectional curvature $K \geq 0$. We outline in this section the construction of a compact, totally convex submanifold S without boundary of M, which will be called a *soul* of M, following [37]. By the results of the previous section, S is totally geodesic, and in particular, also has nonnegative curvature.

We begin by recalling special cases of three classical theorems (for a proof, see [35], [57], or [104]) tailored to our situation. The second one, commonly referred to as the second Rauch comparison theorem, is actually due to Berger.

Theorem 3.2.1 (Rauch I). *Let $p \in M$, and consider a geodesic bi-angle $c_i : [0,1] \to M$, $i = 1, 2$, at p with angle α. If the length of c_i is less than the injectivity radius at p, then the distance between $c_1(1)$ and $c_2(1)$ is no larger than the distance between the endpoints of the bi-angle with same lengths and angle in $\mathbb{R}^2$.*

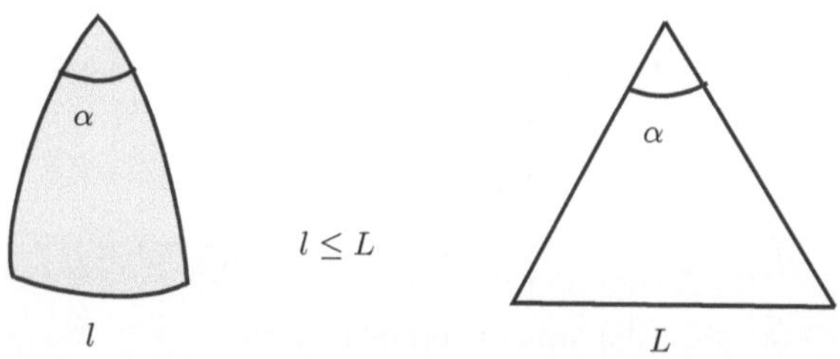

Theorem 3.2.2 (Rauch II). *Let $c : [0, a] \to M$ be a normal geodesic, X a parallel vector field along c, and denote by $\gamma : [0, a] \to M$ the curve given by $\gamma(t) = \exp_{c(t)} X(t)$. If none of the geodesics $s \mapsto \exp sX(t)$ has focal points in $(0, 1)$, then*

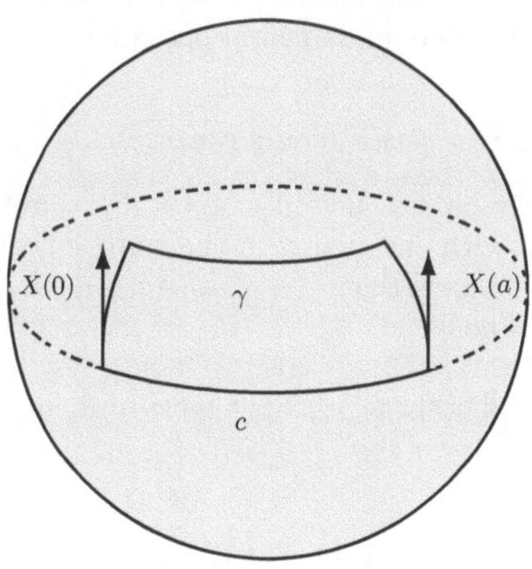

the length $L(\gamma)$ of γ is no larger than a. Furthermore, if it equals a, then the "rectangle"

$$\begin{aligned} V : [0,a] \times [0,1] &\to M, \\ (t,s) &\mapsto \exp_{c(t)} sX(t) \end{aligned}$$

is flat and totally geodesic.

Theorem 3.2.3 (Toponogov). *Let c_i denote the sides of a geodesic triangle in M with angle α_i at the vertex opposite $\dot{c}_i$, $i = 0, 1, 2$. Suppose that the geodesics c_1 and c_2 are minimal, with $L(c_1) + L(c_2) \geq L(c_0)$. Then there exists a triangle in $\mathbb{R}^2$ with sides $\bar{c}_i$ and angles $\bar{\alpha}_i$, such that $L(c_i) = L(\bar{c}_i)$ for all i, and $\alpha_i \geq \bar{\alpha}_i$ for $i = 1, 2$.*

Recall that a *ray* in a noncompact Riemannian manifold M is a geodesic $c : [0, \infty) \to M$ such that $d(c(0), c(t)) = t$ for all $t \geq 0$. When M is complete, there exists at least one ray emanating from any point p of M: To see this, consider a sequence q_n of points in M with $d(p, q_n) \to \infty$. By completeness, there is a minimal normal geodesic c_n from p to q_n. The sequence $\dot{c}_n(0)$ in the unit sphere of M_p must subconverge to some v. The geodesic $t \mapsto tv$ is then a ray, because the function

$$\begin{aligned} s : \{v \in M_p \mid |v| = 1\} &\to \mathbb{R}^+ \cup \{\infty\}, \\ v &\mapsto \sup\{t > 0 \mid d(p, \exp(tv)) = t\} \end{aligned}$$

is continuous, so that $s(\dot{c}_n(0))$ subconverges to $s(v)$.

Now fix a point p in M, a ray c emanating from p, and define

$$B_c = \bigcup_{t>0} B_t(c(t)).$$

Notice that this is an expanding union, since $B_{t_1}(c(t_1)) \subset B_{t_2}(c(t_2))$ for $t_1 < t_2$ by the triangle inequality. The fundamental ingredient in the soul construction is given by the following:

Theorem 3.2.4. *$M \setminus B_c$ is a closed totally convex set.*

Proof. $M \setminus B_c$ is clearly closed. If it is not totally convex, then there exists a geodesic $\gamma : [0, 1] \to M$ with end points in the complement of B_c, but $\gamma(s) \in B_c$ for some $s \in (0, 1)$. It follows that $\gamma(s) \in B_{t_0}(c(t_0))$ for some $t_0 > 0$; set $\epsilon := t_0 - d(\gamma(s), c(t_0)) > 0$. Then

$$d(\gamma(s), c(t)) \leq t - \epsilon \text{ for all } t \geq t_0. \tag{3.2.1}$$

Next, fix some t such that

$$t > \max\{t_0, L(\gamma), L^2(\gamma)/\epsilon\}, \tag{3.2.2}$$

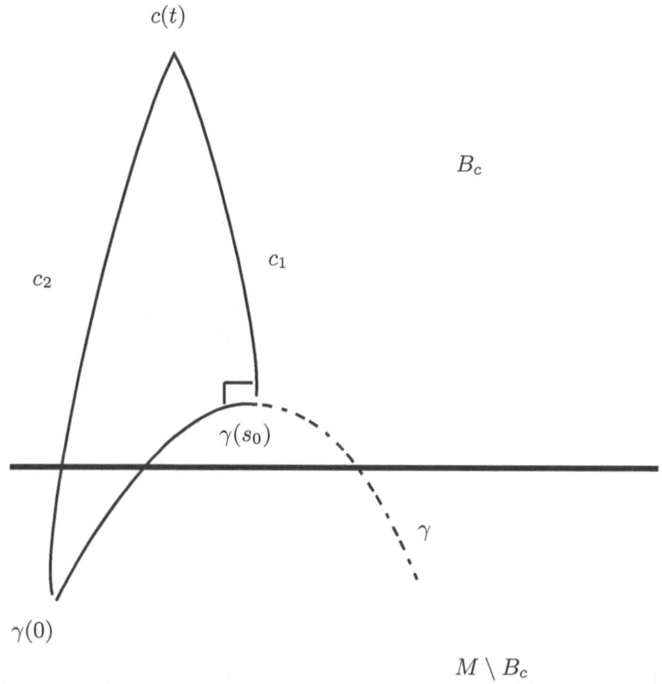

and consider a point $\gamma(s_0)$ on γ that is closest to $c(t)$. Denote by $c_0 := \gamma_{|[0,s_0]}$ the restriction of γ to $[0, s_0]$, and by c_1, c_2 minimal geodesics from $c(t)$ to $\gamma(s_0)$, $\gamma(0)$, respectively. Since $c_0(0) \notin B_t(c(t))$, $L(c_2) > t$, so that $L(c_1)+L(c_2) > t > L(\gamma) > L(c_0)$, and there exists, by Theorem 3.2.3, a comparison triangle in Euclidean space with $\bar{\alpha}_2 \le \alpha_2 = \pi/2$ (the last equality holds because $c_0(s_0)$ is the point on γ closest to $c(t)$ and $s_0 \in (0,1)$). On the other hand, (3.2.1) implies

$$L(c_1) < L(c_2) - \epsilon,$$

so that by the law of cosines in Euclidean space,

$$\begin{aligned}\cos\bar{\alpha}_2 &= \frac{L^2(c_0)+L^2(c_1)-L^2(c_2)}{2L(c_0)L(c_1)}\\ &= \frac{L(c_1)+L(c_2)}{2L(c_1)}\cdot\frac{L(c_1)-L(c_2)}{L(c_0)}+\frac{L(c_0)}{2L(c_1)}\\ &< \frac{1}{2L(c_1)}(L(c_0)-\epsilon\frac{L(c_1)+L(c_2)}{L(c_0)})\\ &< \frac{1}{2L(c_1)}(L(c_0)-\frac{\epsilon t}{L(c_0)}) < 0,\end{aligned}$$

since by (3.2.2), $L^2(c_0) < L^2(\gamma) \le \epsilon t$. This contradicts $\bar{\alpha}_2 \le \pi/2$. □

Even though $M \setminus B_c$ is not, in general, compact, it is now easy to construct a compact totally convex set:

Proposition 3.2.1. *For any $p \in M$, there exists a compact totally convex set C_0 with $p \in \partial C_0$.*

Proof. Define

$$C_0 = \bigcap \{M \setminus B_c \mid c \text{ is a ray }, c(0) = p\}.$$

C_0 is clearly closed, totally convex, and p belongs to its boundary. It remains to establish compactness. If C_0 is not compact, then there exists a sequence of points $p_n \in C_0$ with $d(p, p_n) \to \infty$. Letting c_n denote the minimal normal geodesic in C_0 joining p to p_n, it follows that there exists a subsequence of $\dot{c}_n(0)$ converging to some unit vector $v \in M_p$. By construction, the geodesic $t \mapsto \exp(tv)$ is a ray contained in C_0, contradicting the definition of that set. □

Remark 3.2.1. For any given ray c emanating from p, and $t > 0$, let c_t denote the ray given by $c_t(s) := c(s + t)$. If we define

$$C_t = \bigcap \{M \setminus B_{c_t} \mid c \text{ is a ray }, c(0) = p\},$$

then the same argument as above shows that C_t is a compact totally convex set, and M equals the expanding union $\cup_{t \geq 0} C_t$. It is, furthermore, not difficult to see that if $t_2 \geq t_1$, then

$$C_{t_1} = \{q \in C_{t_2} \mid d(q, \partial C_{t_2}) \geq t_2 - t_1\}.$$

Our next aim is to gradually contract the set C_0 from Proposition 3.2.1 without losing total convexity. This can be achieved with the following:

Theorem 3.2.5. *Let C be a closed totally convex set with boundary in M. Then the distance function*

$$\begin{aligned} f : C &\to \mathbb{R}, \\ q &\mapsto d(q, \partial C) \end{aligned}$$

to the boundary is concave. Furthermore, suppose that for a normal geodesic c in C, the restriction of $f \circ c$ is a constant d on some interval $[a, b]$, and consider the parallel vector field X along c, where $t \mapsto \exp tX(a)$ denotes any minimal normal geodesic from $c(a)$ to ∂C. Then for any $s \in [a, b]$, $t \mapsto \exp tX(s)$ is a minimal geodesic of length d from $c(s)$ to ∂C, and the rectangle

$$\begin{aligned} V : [a, b] \times [0, d] &\to C, \\ (s, t) &\mapsto \exp_{c(s)} tX(s) \end{aligned}$$

is flat and totally geodesic.

Proof. Let $c : [\alpha, \beta] \to C$ be a normal geodesic. In order to establish concavity of $f \circ c$, it suffices to show that for any $s_0 \in (\alpha, \beta)$, $f \circ c$ is bounded above on a neighborhood of s_0 by the linear function $s \mapsto (f \circ c)(s_0) - (\cos \phi)(s - s_0)$, where ϕ denotes the angle between c and the minimal normal geodesic c_{s_0} from $c(s_0)$ to ∂C. We actually only need to do this for $s > s_0$, since the case $s < s_0$ follows by considering $f \circ \tilde{c}$, where $\tilde{c}$ denotes c with the reverse parametrization $\tilde{c}(s) = c(\alpha + \beta - s)$. So set $d := (f \circ c)(s_0)$, and suppose first that $\phi = \pi/2$. Denote by X the parallel field along c_{s_0} with $X(0) = \dot{c}(s_0)$, and consider the rectangle

$$\begin{aligned} V : [s_0, \beta] \times [0, d] &\to M, \\ (s, t) &\mapsto \exp_{c_{s_0}(t)}(s - s_0)X(t). \end{aligned}$$

By Theorem 3.2.2, for s close enough to s_0, each curve $t \mapsto V(s, t)$ has length $\leq d$, and connects $c(s)$ to a point that does not belong to the interior of C, by Proposition 3.1.1. The claim then clearly follows, as does the second statement of the theorem (by the rigidity part of Rauch II). Next, consider the case $\phi > \pi/2$. Denote by v the unit vector that is the convex combination of $\dot{c}_{s_0}(0)$ and $\dot{c}(s_0)$ which is orthogonal to $\dot{c}_{s_0}(0)$. Just as above, it follows that for small t,

$$d(\exp tv, \partial C) \leq d. \tag{3.2.3}$$

On the other hand, by Theorem 3.2.1 and the law of cosines in the plane,

$$d^2(\exp tv, c(s)) \leq t^2 + (s - s_0)^2 - 2t(s - s_0) \cos\left(\phi - \frac{\pi}{2}\right)$$

for small t, and s close enough to s_0. In particular, letting $t = (s - s_0)\cos(\phi - \pi/2)$, we obtain

$$d(\exp tv, c(s)) \leq (s - s_0) \sin\left(\phi - \frac{\pi}{2}\right) = -\cos\phi(s - s_0). \tag{3.2.4}$$

(3.2.3) and (3.2.4), together with the triangle inequality, then establish the claim. Finally, suppose $\phi < \pi/2$, and denote by a_s the minimal normal geodesic from c_{s_0} to $c(s)$, with $a_s(0) = c_{s_0}(t_s)$. Then, as before,

$$d(c(s), \partial C) \leq d - t_s, \tag{3.2.5}$$

and

$$d^2(c_{s_0}(t_s), c(s)) \leq t_s^2 + (s - s_0)^2 - 2t_s(s - s_0)\cos\phi.$$

On the other hand, $\dot{a}_s(0) \perp \dot{c}_{s_0}(t_s)$ because a_s is minimal, so that

$$(s - s_0)^2 \leq d^2(c_{s_0}(t_s), c(s)) + t_s^2.$$

Together, these two inequalities imply that $2t_s(s - s_0)\cos\phi \leq 2t_s^2$, or $t_s \geq (s - s_0)\cos\phi$. Substituting this in (3.2.5) once again establishes the claim. □

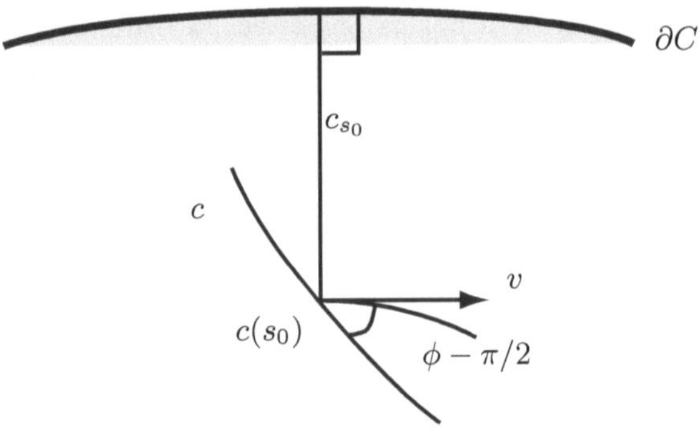

Figure 3.1: The case $\phi > \pi/2$

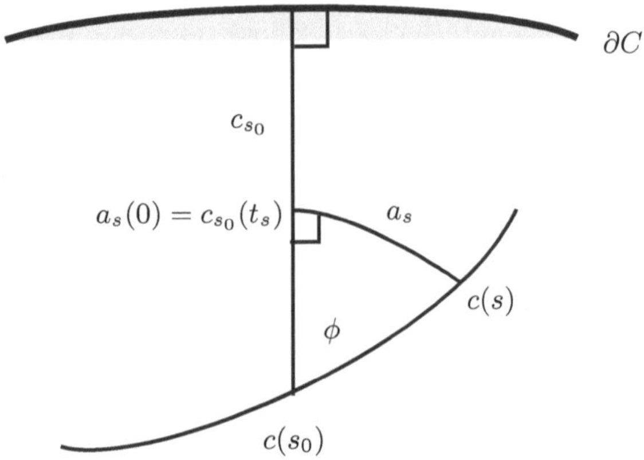

Figure 3.2: The case $\phi < \pi/2$

For a closed totally convex set C_0 with boundary, and $\alpha \geq 0$, define

$$C_0^\alpha = \{q \in C_0 \mid d(q, \partial C_0) \geq \alpha\}, \qquad C_1 = \bigcap \{C_0^\alpha \mid C_0^\alpha \neq \emptyset\}.$$

Theorem 3.2.5 then immediately implies:

Corollary 3.2.1. *C_0^α and C_1 are totally convex, and* $\dim C_1 < \dim C_0$.

Now choose a point p in M, and consider the compact totally convex set C_0 with boundary from Proposition 3.2.1. If C_1 has nonempty boundary, repeat the above procedure finitely many times to conclude:

Theorem 3.2.6. *M contains a compact, totally geodesic submanifold S without boundary.*

In fact, S is totally convex. A submanifold S obtained by this construction is called a *soul* of M.

Examples 3.2.1. (i) If $M = \mathbb{R}^n$ with the standard flat metric, then the set C_0 from Proposition 3.2.1 constructed at some point p consists precisely of p, since every geodesic from p is a ray. Thus, any point of $\mathbb{R}^n$ is a soul.

(ii) Let $M = \{(x, y, z) \in \mathbb{R}^3 \mid z = x^2 + y^2\}$ together with the metric of positive curvature induced from $\mathbb{R}^3$. Given $0 \neq a \in \mathbb{R}^2$, there is exactly one ray in M emanating from $p = (a, |a|^2) \in M$, namely the one whose image consists of the upward branch of the meridian through p. It follows that C_0 consists of those points of M at height $\leq |a|^2$, and the (unique) soul C_1 consists of the vertex of the paraboloid.

(iii) Let $M = S^1 \times \mathbb{R}$ with the product metric. Given $p = (z, t_0) \in M$, there are two rays emanating from p, namely $t \mapsto (z, t_0 + t)$, and $t \mapsto (z, t_0 - t)$. This implies that $C_0 = S^1 \times \{t_0\}$ is a soul of M for any $t_0 \in \mathbb{R}$.

(iv) Let $M = G/H$ be an n-dimensional Riemannian homogeneous space of nonnegative curvature, $p := eH$. By Example 2.7.1, the tangent bundle TM of M is canonically identified with $G \times_H M_p$, and there exists a unique metric of nonnegative curvature on TM for which the projection $\rho : G \times M_p \to TM = G \times_H M_p$ becomes a Riemannian submersion. Given $g \in G$, and a unit $u \in M_p$, the geodesic $t \mapsto (g, tu)$ in $G \times M_p$ is ρ-horizontal, and therefore projects to a geodesic $c_{g,u}$ in TM with $\dot{c}_{g,u}(0) \perp \rho(G \times \{0\})$. We will identify the zero section $\rho(G \times \{0\}) = G \times_H \{0\}$ of the bundle $G \times_H M_p \to G \times_H \{0\}$ with M, so that $p = \rho(e, 0)$. Now, the distance between $c_{g,u}(0)$ and $c_{g,u}(t)$ equals the distance between their pre-images; i.e., between the sets $gH \times \{0\}$ and $\{(gh^{-1}, th_* u) \mid h \in H\}$ in $G \times M_p$. Since the latter is clearly t, $c_{g,u}$ is a ray in TM. But every normal geodesic orthogonal to M is of this form, so that every such geodesic is a ray. Now perform the soul construction at p. We claim that the soul is then M, and coincides with C_0, the first totally convex set from Proposition 3.2.1. To see this, consider any $q \notin M$, and a minimal geodesic $\gamma : [0, a] \to TM$ from q to p. By assumption, there exists a unit vector $u \perp M_p$ that makes an angle $< \pi/2$ with $-\dot{\gamma}(a)$. Since the geodesic $t \mapsto c_u(t) := \exp tu$ is a ray, $q \in B_{c_u}$, and therefore $q \notin C_0$. Thus, $C_0 \subset M$. Conversely, suppose $q \in M$. Notice that M is totally geodesic in TM (if γ is a horizontal geodesic for the submersion $G \to G/H$, then $(\gamma, 0)$ is a horizontal geodesic for ρ and thus projects to a geodesic in M). So consider a geodesic c from p to q. Since $c(\mathbb{R})$ is contained in the compact set M, it is contained in one of the totally convex sets C_t that provide the expanding filtration of M described in Remark 3.2.1. Then the function $s \mapsto d(c(s), \partial C_t)$ is concave, bounded below by 0, and defined on all of $\mathbb{R}$, so that it must be constant equal to $d(p, \partial C_t)$. But $p \in \partial C_0$, and thus, $d(q, \partial C_t) = d(p, \partial C_t) = t$; i.e., $q \in C_0$. Summarizing, we have that $M = C_0$, and since M has empty boundary, it must be a soul, as claimed.

Notice that in the above examples, we have Riemannian submersions onto the soul. We will later see that this is true in general.

(v) Let c be a ray in M. For fixed $p \in M$, the function $t \mapsto d(p, c(t)) - t$ is bounded in absolute value, since

$$|d(p, c(t)) - t| = |d(p, c(t)) - d(c(0), c(t))| \leq d(p, c(0)).$$

For $t \geq 0$, define a function b_t on M by $b_t(q) := d(q, c(t)) - t$. It follows again from the triangle inequality that if $s < t$, then $b_s(q) \geq b_t(q)$. Furthermore, b_t is distance-decreasing in the sense that

$$|b_t(p) - b_t(q)| = |d(p, c(t)) - d(q, c(t))| \leq d(p, q), \qquad p, q \in M.$$

In particular, the family of functions $\{b_t\}_{t \geq 0}$ is equicontinuous and converges to a continuous function b_c, where

$$b_c(p) := \lim_{t \to \infty} d(p, c(t)) - t, \qquad p \in M,$$

which satisfies $|b_c(p) - b_c(q)| \leq d(p, q)|$ for any p, $q \in M$. b_c is called the *Busemann function* of c. The level sets $b_c^{-1}(t)$, $t \geq 0$, are called *horospheres*, and the sets $b_c^{-1}(-\infty, t)$ *horoballs* . They can be thought of as spheres and balls centered at infinity: in fact, the closed half-space $M \setminus B_c$ from Theorem 3.2.4 is just $b_c^{-1}[0, \infty)$. Notice that Theorem 3.2.5 implies that Busemann functions are concave when the ambient space is nonnegatively curved.

3.3 The topological structure of M

Let S denote a soul of M, and $\nu(S) = \{u \in M_p \mid p \in S, u \perp S_p\}$ the total space of the normal bundle of S in M. The aim of this section is to show that M is diffeomorphic to $\nu(S)$. It should be noted, however, that this identification is not canonical; in particular, the exponential map $\nu(S) \to M$ of the normal bundle is not, in general, a diffeomorphism. Instead, we use a generalized notion of regular points of the distance function that dates back to Grove and Shiohama [66].

Let $\rho : M \to \mathbb{R}$ denote the distance function from S. By compactness of S, there exists $\epsilon > 0$ such that the exponential map, when restricted to the disk bundle $\nu^\epsilon(S)$ of vectors of length $< \epsilon$, is a diffeomorphism onto $B_\epsilon(S)$. Thus, on $B_\epsilon(S) \setminus S$, ρ has no critical points, and its gradient $\nabla\rho$ is a vector field of unit length. We begin by generalizing the concept of a regular point of ρ:

Definition 3.3.1. $q \in M \setminus S$ is said to be a *regular point* of ρ if there exists $v \in M_q$ such that for any minimal geodesic c from q to S,

$$\measuredangle(v, \dot{c}(0)) > \frac{\pi}{2}.$$

Such a vector v is called a *gradient-like* vector.

Clearly, a regular point in the usual sense is also regular in the sense of Definition 3.3.1.

Lemma 3.3.1. *Any $q \in M \setminus S$ is a regular point of ρ.*

Proof. By hypothesis, q belongs to the boundary of some totally convex set C that contains S. By the discussion following Proposition 3.1.1, $C_q \setminus \{0\}$ is contained in an open half-space $\{v \in M_q \mid \measuredangle(v,u) < \pi/2\}$ for some $0 \neq u \in M_q$. $-u$ is then a gradient-like vector. $\square$

Theorem 3.3.1. *M is diffeomorphic to the normal bundle $\nu(S)$ of S in M.*

Proof. For each $p \in M \setminus B_{\epsilon/2}(S)$, choose a gradient-like vector v^p as in the lemma, and extend it locally to a vector field X^p. X^p is still gradient-like on a neighborhood U_p of p, which we assume to be contained inside the ball of radius $\epsilon/4$ around p. Let $\{U_i\}_{i\geq 1}$ be a locally finite subcover, with $U_i = U_{p_i}$, and set $U_0 = B_\epsilon(S)$. If $\{\phi_i\}$ is a partition of unity subordinate to $\{U_i\}_{i\geq 0}$, define vector fields X_i on $M \setminus S$ by

$$X_i(p) = \begin{cases} \phi_i(p)X^{p_i}(p), & p \in U_i, \\ 0, & p \notin U_i, \end{cases}$$

for $i \geq 1$, and

$$X_0(p) = \begin{cases} \phi_0(p)\nabla\rho(p), & p \in U_0, \\ 0, & p \notin U_0. \end{cases}$$

Then $X := (\sum_i X_i)/|\sum_i X_i|$ is a smooth gradient-like vector field on $M \setminus S$ that equals $\nabla\rho$ when restricted to $B_{\epsilon/4}(S)$. In particular, ρ is strictly increasing along any integral curve of X. Furthermore, if γ_p denotes the maximal integral curve of X with $\gamma_p(0) = p \in M \setminus S$, there exists $T \in \mathbb{R}$ such that $\gamma_p(T) \in \partial B_{\epsilon/4}(S)$: this is clear for $p \in B_{\epsilon/4}(S)$, since $X = \nabla\rho$ on this set. For $p \notin B_{\epsilon/4}(S)$, let $\gamma(t) = \gamma_p(-t)$. If γ does not reach $\partial B_{\epsilon/4}(S)$ in finite time, then γ is defined on $[0,\infty)$ (having image inside a compact set). But then a subsequence $\gamma_i := \gamma_{|[i,i+1]}$ would converge to an integral curve of X along which ρ is constant, which is impossible. Thus, there exists a differentiable function $T : M \setminus S \to \mathbb{R}$ defined by the condition

$$\gamma_p(T(p)) \in \partial B_{\epsilon/4}(S).$$

Next, let $\psi : [0,\infty) \to [0,\epsilon)$ be a smooth function with positive derivative, that equals the identity when restricted to $[0,\epsilon/4]$. If Φ denotes the maximal flow of X, $\Phi(t,p) = \gamma_p(t)$, define

$$F : M \to B_\epsilon(S),$$
$$p \mapsto \begin{cases} p, & p \in S, \\ \Phi(\psi(\frac{\epsilon}{4} - T(p)) + T(p) - \frac{\epsilon}{4}, p), & p \notin S. \end{cases}$$

Notice that for $p \in B_{\epsilon/4}(S)$, $0 \leq T(p) \leq \epsilon/4$, so that $0 \leq (\epsilon/4) - T(p) \leq \epsilon/4$, and $F(p) = p$. Thus, F is differentiable, and is in fact a diffeomorphism. Since $\nu(S)$ in turn is diffeomorphic to $B_\epsilon(S)$, this completes the proof. $\square$

3.4 The Sharafutdinov retraction

Our next project is the construction of a continuous, distance non-increasing map $\pi : M \to S$ onto a soul of M; i.e., $d(\pi(p), \pi(q)) \le d(p, q)$ for p, $q \in M$. In fact, a stronger result will be established, namely: There exists a strong deformation retraction $H : M \times [0,1] \to M$ onto S such that the map $p \mapsto H(p,t)$ is distance non-increasing for every $t \in [0,1]$. π is then the map corresponding to $t = 1$. The construction of such a map is due to Sharafutdinov [115]. Here we follow an alternative approach by Schroeder and Ziller, cf. [144], [145]. Although the argument in itself doesn't preclude the existence of other distance non-increasing retractions onto a soul, we will see in the next section that such a map is unique.

A key ingredient in the proof is the following:

Lemma 3.4.1 ([30], 6.3.5). *Let c be a normal geodesic in a Riemannian manifold with sectional curvature $K \le \kappa$, and J a Jacobi field orthogonal to $\dot c$. Denote by c_κ a normal geodesic in the simply connected two-dimensional space of constant curvature κ, and by J_κ the Jacobi field orthogonal to $\dot c_\kappa$ with the same initial conditions as J: $|J_\kappa|(0) = |J|(0)$, $|J_\kappa|'(0) = |J|'(0)$. If $J_\kappa \neq 0$ on $[0,a)$, then $|J| \ge |J_\kappa|$ on $[0,a]$.*

Corollary 3.4.1. *Let $\gamma : [0,l] \to M$ be a normal geodesic in a Riemannian manifold M with curvature bounded above by $\kappa > 0$, and X a vector field along γ orthogonal to $\dot\gamma$ which is contained in the domain of $\exp|_{\gamma[0,l]}$. Given a smooth surjective function $\phi : [0,1] \to [0,l]$, define a curve $c : [0,1] \to M$ by*

$$c(t) = \exp_{(\gamma\circ\phi)(t)}(X \circ \phi)(t).$$

If $|X| \le \epsilon < \pi/(2\sqrt{\kappa})$, then $L(c) \ge \cos(\epsilon\sqrt{\kappa}) \cdot L(\gamma)$.

Proof. The result will follow once we establish that

$$|\dot c|(t) \ge |\widehat{\gamma\circ\phi}^{\cdot}|(t)\cos(\epsilon\sqrt{\kappa}), \qquad t \in [0,1], \tag{3.4.1}$$

since $L(c) = \int_0^1 |\dot c|$, and $L(\gamma) \le \int_0^1 |\dot{\widehat{\gamma\circ\phi}}|$ by surjectivity of ϕ:

$$\int_0^1 |\dot{\widehat{\gamma\circ\phi}}| = \int_0^1 |\phi'(\dot\gamma\circ\phi)| = \int_0^1 |\phi'| \ge l = L(\gamma).$$

So set $X_0 := X/|X|$, and consider the variation $V : [0,1]\times[0,1] \to M$ given by

$$V(t,s) = \exp_{(\gamma\circ\phi)(t)} s(X_0\circ\phi)(t).$$

For any fixed $t_0 \in [0,1]$, the curve c_{t_0}, where $c_{t_0}(s) = V(t_0, s)$, is a geodesic, and the vector field J along c_{t_0}, where $J(s) = V_* D_1(t_0, s)$, is Jacobi. By definition,

$$J(0) = \dot{\widehat{\gamma\circ\phi}}(t_0) \perp \dot c_{t_0}(0) = (X_0\circ\phi)(t_0). \tag{3.4.2}$$

Furthermore,

$$\begin{aligned}\langle J, \dot{c}_{t_0}\rangle'(0) &= \langle \nabla_{D_2} V_* D_1, V_* D_2\rangle(t_0,0) = \langle \nabla_{D_1} V_* D_2, V_* D_2\rangle(t_0,0)\\ &= \frac{1}{2}|X_0 \circ \phi|'^2(t_0) = 0.\end{aligned}$$

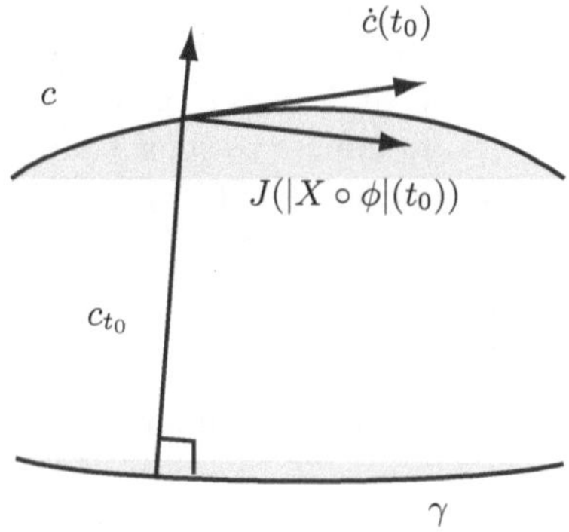

Together with (3.4.2), this implies

$$J \perp \dot{c}_{t_0}. \tag{3.4.3}$$

Now, $c(t) = V(t, |X \circ \phi|(t))$, so that

$$\begin{aligned}\dot{c}(t_0) &= V_* D_1(t_0, |X \circ \phi|(t_0)) + |X \circ \phi|'(t_0) V_* D_2(t_0, |X \circ \phi|(t_0))\\ &= J(|X \circ \phi|(t_0)) + |X \circ \phi|'(t_0)\dot{c}_{t_0}(|X \circ \phi|(t_0)),\end{aligned}$$

and by (3.4.3),

$$|\dot{c}|(t_0) \geq |J|(|X \circ \phi|(t_0)). \tag{3.4.4}$$

On the other hand,

$$\begin{aligned}\langle J, J'\rangle(0) &= \langle \nabla_{D_2} V_* D_1, V_* D_1\rangle(t_0,0) = \langle \nabla_{D_1} V_* D_2, V_* D_1\rangle(t_0,0)\\ &= D_1\langle V_* D_2, V_* D_1\rangle(t_0,0) - \langle V_* D_2, \nabla_{D_1} V_* D_1\rangle(t_0,0)\\ &= \langle X_0 \circ \phi, \widehat{\dot{\gamma} \circ \phi}\rangle'(t_0) - \langle X_0 \circ \phi, \nabla_D \widehat{\dot{\gamma} \circ \phi}\rangle(t_0).\end{aligned}$$

The first term vanishes because $X_0 \perp \dot{\gamma}$. So does the second one, since

$$\nabla_D \widehat{\dot{\gamma} \circ \phi} = \nabla_D \phi'(\dot{\gamma} \circ \phi) = \phi''(\dot{\gamma} \circ \phi) \perp X_0 \circ \phi.$$

Thus, $|J|'(0) = 0$, and by Lemma 3.4.1, $|J|(t) \geq |J|(0)\cos(t\sqrt{\kappa})$. Together with (3.4.2) and (3.4.4), this yields

$$\begin{aligned}|\dot{c}|(t_0) &\geq |J|(0)\cos(|X \circ \phi|(t_0) \cdot \sqrt{\kappa}) \geq |J|(0)\cos(\epsilon\sqrt{\kappa})\\ &= |\widehat{\dot{\gamma} \circ \phi}|(t_0)\cos(\epsilon\sqrt{\kappa}),\end{aligned}$$

which proves (3.4.1). □

Theorem 3.4.1. *Let M be an open manifold of nonnegative curvature with soul S. There exists a strong deformation retraction $H : M \times [0,1] \to M$ onto S, such that if $\imath_t : M \to M \times [0,1]$ is the imbedding $\imath_t(p) = (p,t)$, then $H \circ \imath_t$ is distance non-increasing for each $t \in [0,1]$.*

Proof. Recall from Section 3.2 that there exists a family of compact totally convex sets such that $M = \cup_{t \geq 0} C^t$, $C^0 =: C_0 \supset C_1 \supset \cdots \supset C_k = S$, where C_{i+1} denotes the set of points at maximal distance from the boundary of C_i. Let C be a compact totally convex set with boundary, and for $\alpha \geq 0$, $C^\alpha = \{p \in C \mid d(p, \partial C) \geq \alpha\}$. Set $\alpha_0 = \sup\{d(p, \partial C) \mid p \in C\}$. It suffices to show that for $0 \leq \alpha \leq \beta \leq \alpha_0$, there exists a strong deformation retraction $H_\alpha^\beta : C^\alpha \times [0,1] \to C^\alpha$ onto C^β such that $H_\alpha^\beta \circ \imath_t$ is distance non-increasing for each t.

We begin by assuming that $\beta < \alpha_0$. Since C is compact, there exists ϵ_0 such that for $p \in C$ and $0 < r \leq \epsilon_0$,

Property 1: $B_r(p)$ is the strictly convex diffeomorphic image of the Euclidean ball of radius r about $0 \in M_p$ under $\exp_p$.

Property 2: If $c : [0,l] \to B_r(p)$ is a nonconstant geodesic, $\gamma : [0,1] \to B_r(p)$ the minimal geodesic from p to $c(0)$, and $\measuredangle(\dot{c}(0), \dot{\gamma}(1)) < \pi/2$, then $t \mapsto d(p, c(t))$ is strictly increasing along $[0,l]$.

We will assume that $\epsilon_0 < \pi/(2\sqrt{\kappa})$, where κ is an upper bound for the sectional curvatures of planes in TC. Set $\epsilon := \epsilon_0/3$. By uniform continuity of the function $(t,p) \mapsto d(p, C^t)$ on the compact set $[0, \alpha_0] \times C$, there exists $\delta > 0$ such that for $p \in C^t$, $d(p, C^s) < \epsilon$ if $0 < s - t < \delta$. We may assume that $\beta - \alpha < \delta$, since $[\alpha, \beta]$ can be divided into subintervals $[t_{i-1}, t_i]$ of length $< \delta$, and the deformation retraction H_α^β may then be constructed by composing the successive retractions $H_{t_{i-1}}^{t_i}$. Next, observe that there is a well-defined projection

$$h : C^\alpha \to C^\beta, \qquad \text{where } d(p, h(p)) = d(p, C^\beta).$$

In fact, if there exist two points q_1, $q_2 \in C^\beta$ such that $d(p, q_1) = d(p, q_2) = d(p, C^\beta)$, then the minimal geodesic $c : [0,1] \to C^\beta$ from q_1 to q_2 is contained in $B_{\epsilon_0}(p)$. If $\gamma : [0,1] \to C^\alpha$ is the minimal geodesic from p to q_1, then $\measuredangle(\dot{c}(0), \dot{\gamma}(1)) < \pi/2$ (because $c(0) \in C^\beta$ is a point closest to p), and by Property 2, $t \mapsto d(p, c(t))$ is strictly increasing, contradicting the fact that this function takes on the same value at 0 and 1.

So consider points p, $q \in C^\alpha$ at distance less than ϵ apart. Then $d(p, h(p))$ and $d(q, h(q))$ are less than ϵ, and q, $h(p)$, and $h(q)$ are all contained in the strictly convex ball $B_{\epsilon_0}(p)$. Denote by $c : [0,a] \to C^a$, $\gamma : [0,l] \to C^\beta$ the minimal normal geodesics from p to q and from $h(p)$ to $h(q)$ respectively. There is a well-defined continuous metric projection $\pi : c[0,a] \to \gamma[0,l]$, where $d(c(t), \pi(c(t))) = d(c(t), \gamma[0,l])$. π is surjective, since $\pi(c(0)) = \gamma(0)$, $\pi(c(a)) = \gamma(l)$, and $\pi \circ c$ is defined on a connected interval. Let t_0 (respectively t_1) denote the supremum (respectively infimum) of those $t \in [0,l]$ such that $(\pi \circ c)(t) = \gamma(0)$ (respectively $\gamma(l)$). Then the restriction $\pi : c[t_0, t_1] \to \gamma[0, l]$ is also surjective, and

$$c(t) = \exp_{(\gamma\circ\phi)(t)}(X \circ \phi)(t),$$

for some $X \perp \dot\gamma$ with $|X| < 2\epsilon < \epsilon_0$, where ϕ is defined by the equation $\gamma\circ\phi = \pi\circ c$. Since $\epsilon_0 < \pi/(2\sqrt{\kappa})$, Corollary 3.4.1 implies

$$d(h(p), h(q)) \le \frac{1}{\cos(2\epsilon\sqrt{\kappa})} d(p, q). \tag{3.4.5}$$

(3.4.5) actually holds for any p, $q \in C^\alpha$, since if $d(p, q) \ge \epsilon$, then γ can be subdivided into geodesic segments $\gamma|_{[t_{i-1},t_i]}$, $1 \le i \le k$, of length less than ϵ, and the above argument may be applied to each segment, so that

$$\begin{aligned} d(h(p), h(q)) &\le \sum_{i=1}^{k} d(h(\gamma(t_{i-1}), h(\gamma(t_i)) \le \frac{1}{\cos(2\epsilon\sqrt{\kappa})} \sum_i d(\gamma(t_{i-1}), \gamma(t_i)) \\ &= \frac{1}{\cos(2\epsilon\sqrt{\kappa})} d(p, q). \end{aligned}$$

Thus, $h : C^\alpha \to C^\beta$ is a Lipschitz map with constant $(\cos\kappa_0)^{-1}$, where $\kappa_0 := 2\epsilon\sqrt{\kappa}$.

Set $f_0 := h$, and define f_1 to be the composition of the projections $C^\alpha \to C^{\alpha+(\beta-\alpha)/2}$ and $C^{\alpha+(\beta-\alpha)/2} \to C^\beta$. Each projection is Lipschitz with constant $(\cos(\kappa_0/2))^{-1}$, so that f_1 has Lipschitz constant $(\cos(\kappa_0/2))^{-2}$. One obtains inductively a sequence $f_i : C^\alpha \to C^\beta$ of equicontinuous maps with constants $(\cos(2^{-i}\kappa_0))^{-2i}$. Since C is compact, we may, by the Ascoli theorem, assume that f_i converges to a continuous map $f : C^\alpha \to C^\beta$, which is distance non-increasing since $(\cos(2^{-i}\kappa_0))^{-2i} \to 1$. Furthermore, each f_i may be expressed as a composition $f_i = f_i^1 \circ f_i^2$, where $f_i^2 : C^\alpha \to C^{\alpha+(\beta-\alpha)/2}$ and $f_i^1 : C^{\alpha+(\beta-\alpha)/2} \to C^\beta$ are Lipschitz maps constructed as above. Since $f_i \to f$, subsequences of f_i^1 and f_i^2 converge to distance non-increasing maps f^1 and f^2, with $f = f^2 \circ f^1$. By an induction argument, f can be expressed for any integer m as a composition

$$f = f^{2^m} \circ f^{2^m-1} \circ \cdots \circ f^1$$

of distance non-increasing maps $f^i : C^{t_{i-1}} \to C^{t_i}$, where $\alpha = t_0 < t_1 < \cdots < t_{2^m} = \beta$ is a partition of $[\alpha, \beta]$ into 2^m intervals of length $(\beta-\alpha)/2^m$. It now follows by a continuity argument that for any $t_0 \in (0, 1)$, $f = f_{t_0}^1 \circ f_{t_0}^2$, where $f_{t_0}^2 : C^\alpha \to C^{\alpha+t_0(\beta-\alpha)}$ and $f_{t_0}^1 : C^{\alpha+t_0(\beta-\alpha)} \to C^\beta$ are distance non-increasing. Then $H_\alpha^\beta : C^\alpha \times [0, 1] \to C^\alpha$, where $H_\alpha^\beta(p, t) = f_t^2(p)$ if $t < 1$ and $f(p)$ if $t = 1$, satisfies the requirements of the theorem.

It remains to consider the case $\beta = \alpha_0$. If $\beta_i \to \alpha_0$ is an increasing sequence, define $H_\alpha^{\beta_i}$ to equal the composition of $H_\alpha^{\beta_{i-1}}$ with a distance non-increasing strong deformation retraction $H_{\beta_{i-1}}^{\beta_i}$ of $C^{\beta_{i-1}}$ onto C^{β_i}, adequately parametrized. We then obtain a homotopy $H : C^\alpha \times [0, 1) \to C^\alpha$ which can be extended to $C^\alpha \times [0, 1]$ by the Ascoli theorem. $\square$

3.5 The metric projection onto the soul

A distance non-increasing retraction $\pi : M \to S$ onto the soul S, such as the one constructed in the last section, need *a priori* not be unique or even differentiable. We now show that in fact any such map π is a C^∞ Riemannian submersion, and is uniquely determined as the *metric projection onto the soul*, in the sense that the diagram

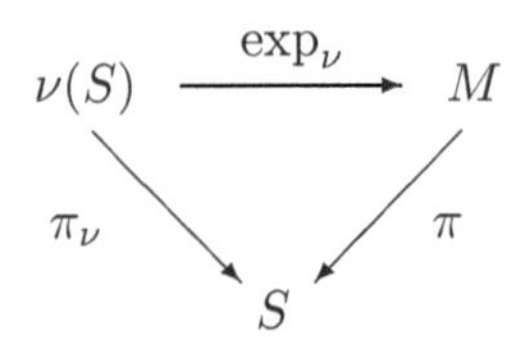

commutes, where $\pi_\nu : \nu(S) \to S$ is the normal bundle projection. Notice that commutativity of the diagram implies that for $p \in M$, $\pi(p)$ is the point of S that is closest to p, so that π is indeed the metric projection. The uniqueness part is implicit in the following theorem, due to Perelman [103]:

Theorem 3.5.1. *If $\pi : M \to S$ is a distance non-increasing retraction, then $\pi \circ \exp_\nu = \pi_\nu$. Furthermore, if $c : \mathbb{R} \to S$ is a geodesic in S, and X a parallel field along c with $X(0) \in \nu(S)$, then the rectangle $V : \mathbb{R} \times [0, \infty) \to M$, $V(t, s) = \exp_{c(t)} sX(t)$, is flat and totally geodesic.*

Proof. For $t \geq 0$, let $\nu^t(S) = \{v \in \nu(S) \mid |v| = t\}$ denote the total space of the sphere bundle of radius t over S, and define $f : [0, \infty) \to \mathbb{R}$ by

$$f(t) := \max\{d(\pi \circ \exp_\nu(v), \pi_\nu(v)) \mid v \in \nu^t(S)\}.$$

For brevity, we drop the subscript ν in $\exp_\nu$, and denote $\pi \circ \exp$ by F. Then $f(0) = 0$, and the first statement will follow once we establish that f is weakly decreasing. The second one is a consequence of the first: Given $t_1, t_2 \in \mathbb{R}$ such that the restriction of c to $[t_1, t_2]$ is minimal and $s_0 > 0$, define $c_{s_0} : [t_1, t_2] \to M$ by $c_{s_0}(t) = \exp_{c(t)} s_0 X(t)$. If s_0 is small enough that none of the geodesics $s \mapsto \exp_{c(t)} sX(t)$ has a focal point in $(0, s_0)$, then by the second Rauch comparison theorem, $L(c_{s_0}) \leq L(c_{|[t_1,t_2]})$. But since π is distance non-increasing,

$$L(c_{|[t_1,t_2]}) = d(c(t_1), c(t_2)) \leq d(c_{s_0}(t_1), c_{s_0}(t_2)) \leq L(c_{s_0}).$$

The rigidity part of Theorem 3.2.2 then implies that the rectangle is flat and totally geodesic up to distance s_0, and in fact up to any distance: consider the set of all $s_0 \geq 0$ for which the rectangle $V : [t_1, t_2] \times [0, s_0] \to M$, $V(t, s) = \exp_{c(t)} sX(t)$, is flat and totally geodesic. It is clearly closed, and by the above argument it is also open. Being nonempty, it equals $[0, \infty)$. Finally, since the domain of c is a union on intervals on which c is minimal, the interval $[t_1, t_2]$ may be replaced by $\mathbb{R}$.

To see that f is decreasing, choose $t \in (0, \mathrm{inj}_S)$, and $v \in \nu^t(S)$ with $f(t) = d(F(v), \pi_\nu(v))$. By assumption, the geodesic in S from $F(v)$ to $p := \pi_\nu(v)$ minimizes past p to some point q. Let w denote the parallel translate of v along this geodesic to q.

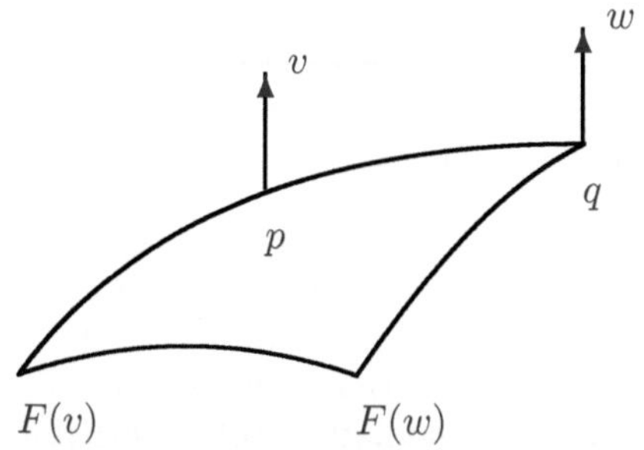

By Rauch II and the fact that π decreases distances,

$$d(F(v), F(w)) \leq d(\exp v, \exp w) \leq d(p, q). \tag{3.5.1}$$

Since $f(t) = d(F(v), p)$,

$$d(F(w), q) \leq f(t). \tag{3.5.2}$$

On the other hand,

$$d(F(v), q) = d(F(v), p) + d(p, q) = f(t) + d(p, q). \tag{3.5.3}$$

Apply the triangle inequality to $F(v)$, $F(w)$, q to conclude that equality holds in (3.5.1) and (3.5.2). By the rigidity part of Rauch II, p, q, $\exp v$, and $\exp w$ determine a flat totally geodesic rectangle.

Next, consider $\epsilon \in (0, t)$, and let $w_\epsilon := (1 - (\epsilon/t))w$, so that $|w_\epsilon| = t - \epsilon$. Then

$$d^2(\exp v, \exp w_\epsilon) \leq d^2(\exp v, \exp w) + \epsilon^2 = d^2(p, q)(1 + (\epsilon^2/d^2(p, q))),$$

so that

$$d(F(v), F(w_\epsilon)) \leq d(\exp v, \exp w_\epsilon) \leq d(p, q) + \frac{\epsilon^2}{2d^2(p, q)} \tag{3.5.4}$$

for small ϵ, by Taylor's theorem. Since $d(F(w_\epsilon), q) \leq f(t - \epsilon)$, (3.5.3) and (3.5.4) imply

$$f(t - \epsilon) \geq d(q, F(v)) - d(F(v), F(w_\epsilon)) \geq f(t) - \frac{\epsilon^2}{2d^2(p, q)}.$$

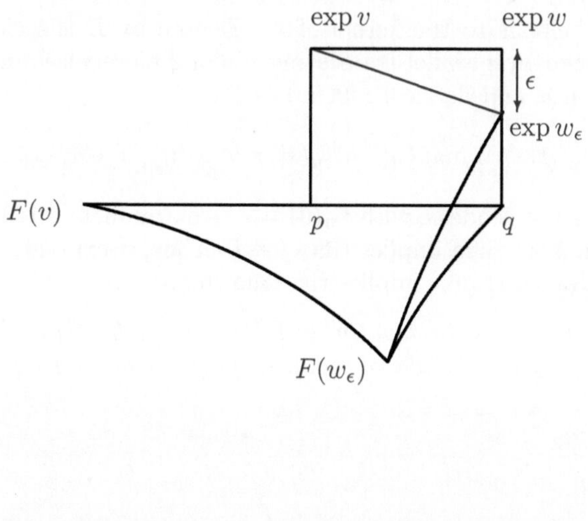

In particular,

$$\limsup_{h \to 0^+} \frac{f(t+h) - f(t)}{h} \leq 0,$$

and f is weakly decreasing, as claimed. More generally, the above argument shows that the set $\{t \geq 0 \mid f(s) = 0 \text{ on } [0,t]\}$ is open, so that f is identically zero. □

Gromoll and Meyer [58] proved that an open manifold of positive curvature is diffeomorphic to $\mathbb{R}^n$. Cheeger and Gromoll [37] conjectured this should still be true under the weaker assumption that the curvature is positive at one point. Perelman's theorem yields an immediate proof of this conjecture:

Corollary 3.5.1. *Let M be an open manifold with nonnegative sectional curvature. If the sectional curvatures at some $p \in M$ are positive, then M is diffeomorphic to Euclidean space.*

Proof. We show the contrapositive statement: If M is not diffeomorphic to Euclidean space (i.e., if a soul S of M does not consist of one point), then S is positive-dimensional, and for any p in M there exist a plane $P \subset M_p$ with $K(P) = 0$. In fact, if $q = \pi(p)$, and $c : [0,1] \to M$ is a minimal geodesic from q to p, then the plane spanned by $\dot{c}(1)$ and the parallel translate along c of any $x \in S_q$ has vanishing curvature by Theorem 3.5.1. □

Next, we tackle the matter of smoothness of π. Observe that since $\pi \circ \exp_\nu = \pi_\nu$, π is C^∞ wherever $\exp_\nu$ is locally 1-1; i.e., Perelman's result implies that π is smooth almost everywhere. Similarly, if $r > 0$ is small enough that $\exp_\nu : \nu(S) \to B_r(S)$ is a diffeomorphism, then $\pi_{|B_r(S)}$ is a C^∞ Riemannian submersion. By the proof of Theorem 3.5.1, any π-horizontal curve remains at constant distance from S, and by the results from Section 1.8, there exists a partition of $B_r(S)$ into dual leaves. The latter are smooth by Section 1.8, and intrinsically complete once again by the proof of Theorem 3.5.1.

Fix one such dual leaf $L \subset B_r(S)$, and consider a horizontal geodesic $c : \mathbb{R} \to L$. The argument used in the proof of Theorem 1.8.1 shows that the normal bundle of L along c is spanned by parallel Jacobi fields, and in particular, must be flat. Thus, if $p \in L$ and $v \in \nu(L) \cap M_p$, then the parallel translate of v along c stays vertical; i.e., it belongs to the kernel of π_*. Denote by E the closure of the set of vectors obtained by parallel translating v along piece-wise smooth horizontal geodesics through p, and define $f : [0, \infty) \to \mathbb{R}$ by

$$f(s) = \max\{d(\pi \circ \gamma_u(s), \pi \circ \gamma_u(0)) \mid u \in E\},$$

where γ_u denotes the geodesic with $\dot{\gamma}_u(0) = u$. An argument virtually identical to that in Theorem 3.5.1 then implies that $f \equiv 0$. Thus, $\pi(\exp(tv)) = \pi(p)$, and the rigidity part of Rauch II now implies the following:

Lemma 3.5.1. *Let L denote a dual leaf in $B_r(S)$, where $B_r(S)$ is the diffeomorphic image via $\exp_\nu$ of the disk bundle of radius r in $\nu(S)$. Consider a piece-wise smooth* *horizontal geodesic* $c : [0,1] \to L$, $p := c(0)$, $q := c(1)$, *and let* X *be a parallel vector field along* c *with* $X(0) \in \nu(L)$. *Then*

1. $\pi(\exp sX(t)) = \pi \circ c(t)$;
2. *The rectangle* $V : [0,1] \times [0,\infty) \to M$, $V(t,s) = \exp_{c(t)} sX(t)$, *is flat and totally geodesic.*
3. *If* $h_c : \pi^{-1}(\pi(p)) \to \pi^{-1}(\pi(q))$ *denotes the holonomy map obtained by horizontal lifts of* $\pi \circ c$, *and* $||_c$ *is parallel translation along* c, *then the diagram*

$$\begin{array}{ccc} \nu_p(L) & \xrightarrow{\ ||_c\ } & \nu_q(L) \\ {\scriptstyle\exp}\downarrow & & \downarrow{\scriptstyle\exp} \\ \pi^{-1}(\pi(p)) & \xrightarrow[h_c]{} & \pi^{-1}(\pi(q)) \end{array}$$

commutes.

The second statement in the lemma essentially says that if $x \in M_p$ is horizontal, and $v \in M_p$ is perpendicular to the holonomy orbit, then x and v generate a totally geodesic flat. The statement actually holds for any metric foliation in a manifold of nonnegative curvature, provided the dual leaves are complete, cf. [141].

A consequence of Perelman's result is that the metric projection π onto a soul is C^1, since the derivative of π is isometric on the continuous horizontal distribution and zero on the vertical one. Guijarro later showed that π is C^2 [68]. Full regularity was established in 2005:

Theorem 3.5.2 (Wilking). *The metric projection* $\pi : M \to S$ *onto the soul is a* C^∞ *Riemannian submersion.*

Proof. In order to show that π is smooth in a neighborhood of $p \in M$, we first prove that the restriction $\pi_{|L}$ of π to the dual leaf L containing p is smooth. So consider a minimal geodesic from p to some dual leaf $L_1 \subset B_r(S)$, where $B_r(S)$ is the diffeomorphic image via exp of a neighborhood of the zero section in the normal bundle of S. Then $p = \exp_{p_1} u$ for some $p_1 \in L_1$ and some $u \in \nu_{p_1}(L_1)$. Let X be the parallel section of $\nu(L_1)$ with $X(p_1) = u$. By Lemma 3.5.1, $L = \{\exp X(z) \mid z \in L_1\}$, and the map

$$\begin{aligned} \phi : L_1 &\to L, \\ z &\mapsto \exp X(z) \end{aligned}$$

is smooth and surjective. Furthermore, if c is a piece-wise smooth horizontal geodesic in L_1 joining p_1 to some q_1, then

$$h_c : L_1 \cap \pi^{-1}(\pi(p_1)) \to L_1 \cap \pi^{-1}(\pi(q_1))$$

is a diffeomorphism, and by Lemma 3.5.1 (3), the diagram

$$\begin{array}{ccc} L_1 \cap \pi^{-1}(\pi(p_1)) & \xrightarrow{h_c} & L_1 \cap \pi^{-1}(\pi(q_1)) \\ {\scriptstyle \phi}\downarrow & & \downarrow{\scriptstyle \phi} \\ L \cap \pi^{-1}(\pi(p_1)) & \xrightarrow[h_c]{} & L \cap \pi^{-1}(\pi(q_1)) \end{array}$$

commutes. This implies that ϕ has constant rank: In fact, the kernel of ϕ_* is vertical (because $\pi \circ \phi = \pi$), so the above diagram says that h_{c*} maps the kernel of ϕ_{*p_1} isomorphically onto the kernel of ϕ_{*q_1}. Thus, ϕ is a smooth submersion. Now, $\pi_{|L} \circ \phi = \pi_{|L_1}$, so that locally there exist charts such that if $\psi_1 : U \subset \mathbb{R}^{l+l_1} \to \mathbb{R}^{l_2}$ represents the smooth map $\pi_{|L_1}$ in this coordinate system, then ψ_1 factors as

$$\begin{array}{ccc} U \subset \mathbb{R}^{l+l_1} & \xrightarrow{\mathsf{p}} & 0 \times \mathbb{R}^{l_1} \\ & {\scriptstyle \psi_1} \searrow \quad \swarrow {\scriptstyle \psi} & \\ & \mathbb{R}^{l_2} & \end{array}$$

where $\mathsf{p} : \mathbb{R}^{l+l_1} \to 0 \times \mathbb{R}^{l_1}$ is projection, and ψ represents $\pi_{|L}$. But then $\psi = \psi_1 \circ \imath$, where $\imath : 0 \times \mathbb{R}^{l_1} \to \mathbb{R}^{l+l_1}$ denotes inclusion, so that ψ, and hence also $\pi_{|L}$, is smooth. To conclude the proof, let $\pi^L : \nu(L) \to L$ denote the bundle projection. By Lemma 3.5.1 (1), $\pi \circ \exp_{\nu(L)} = \pi_{|L} \circ \pi^L$, and since the right side of this identity is smooth, $\pi \circ \exp_{\nu(L)}$ is C^∞. But $\exp_{\nu(L)}$ is a diffeomorphism in a neighborhood of $0 \in \nu_p(L)$, so that π is indeed smooth in a neighborhood of p. Finally, the fact that π is Riemannian was established in Theorem 3.5.1. □

A different approach in attempting to prove the above theorem is adopted in [33].

3.6 The metric structure of bundles with $K \geq 0$

The work of Perelman and Wilking discussed in the last section suggests that vector bundles which admit a metric with nonnegative sectional curvature have a fairly rigid metric structure. The aim of this section is to explore this further. We begin by introducing a note of caution, though: Vector bundles are classified by equivalence classes, whereas existence of metrics with $K \geq 0$ depends only on the diffeomorphism type of the total space. This can lead to surprising facts. For example, there are nontrivial bundles over S^n, discovered by Levine [84], with total space diffeomorphic to $S^n \times \mathbb{R}^k$. Fix one such bundle ξ with total space E, and a diffeomorphism $f : E \to S^n \times \mathbb{R}^k$. Consider the standard product metric on the image space, and pull it back via f to obtain a metric of nonnegative curvature on E. Then E has infinitely many souls (namely $f^{-1}(S^n \times \{u\})$ for any $u \in \mathbb{R}^k$), but the zero section is not a soul, for if it were one, than ξ would be a trivial bundle, see [41] for details. For this reason, we implicitly assume in what follows that a given metric of nonnegative curvature on the total space of a vector bundle has the zero section as soul.

Let M be an open manifold of nonnegative curvature with soul S, E the total space of the normal bundle $\nu(S) = \pi_\nu : E \longrightarrow S$ of S in M. Given $p \in S$, the fiber $\pi_\nu^{-1}(p)$ over p will be denoted by E_p. We begin with the relation between the A-tensor of the submersion, the curvature tensor R^ν of the normal bundle, and the curvature R of the metric on M.

Proposition 3.6.1 ([118]). *Let $p \in S$, x, $y \in S_p$, $u \in E_p$, and c the geodesic $c(t) = \exp(tu)$. If X, Y denote the horizontal lifts of x, y along c, then*

1. $R^\nu(x,y)u = R(x,y)u$,
2. *$A_X Y \circ c$ is the Jacobi field J along c with $J(0) = 0$, $J'(0) = -\frac{1}{2}R(x,y)u$,*
3. $R(x,y)u = 2R(x,u)y$.

Proof. The first identity follows from the fact that S is totally geodesic in M. For the second identity, consider the connection $\tilde{\mathcal{H}}$ on $\nu(S)$ induced by the Levi-Civita connection of M. By Perelman's theorem (Theorem 3.5.1), if U is a parallel section of ν along a curve, then the curve $\exp_\nu U$ is horizontal in M. Thus, $\exp_{\nu *} \tilde{\mathcal{H}} = \mathcal{H} \circ \exp_\nu$, and in the same way, $\exp_\nu$ preserves the vertical subspaces (notice also that $\exp_{\nu *}$ has maximal rank on the horizontal distribution). Given a vector field X on S, denote its horizontal lifts to E and M by $\tilde{X}$ and $\bar{X}$, respectively. Then $\tilde{X}$ and $\bar{X}$ are $\exp_\nu$-related, and by definition of the curvature tensor of the normal bundle, given $u \in E_p$,

$$\mathcal{J}_u(R^\nu(X,Y)u) = -[\tilde{X},\tilde{Y}]^{\mathsf{v}}(u), \qquad X, Y \in \mathfrak{X}(S),$$

where $\mathcal{J}_u$ denotes the canonical isomorphism between the vector space E_p and its tangent space at u, see, e.g., [136]. Applying $\exp_*$ to both sides, we see that if c is a geodesic orthogonal to S, then

$$(2A_{\bar{X}}\bar{Y}) \circ c(t) = [\bar{X},\bar{Y}]^{\mathsf{v}} \circ c(t) = \exp_{\nu *}(-t\mathcal{J}_{t\dot{c}(0)} R^\nu(X,Y)\dot{c}(0)),$$

which proves the second identity. For the last identity, denote by ∂_r the gradient of the distance function to S on a small enough neighborhood of S. By Perelman's result, if X is basic, then $\nabla_X \partial_r = \nabla_{\partial_r} X = 0$, so that

$$R(X,\partial_r)Y = -\nabla_{\partial_r}\nabla_X Y.$$

If u is a unit vector in E, then the curve $t \mapsto c(t) = \exp_\nu(tu)$ is an integral curve of ∂_r, and

$$R(X \circ c, \dot{c})Y \circ c = -J', \qquad J := (A_X Y) \circ c.$$

Evaluating this at $t = 0$ and using (2) now proves (3). □

The proposition also gives an easy proof of a result originally due to Strake [117]:

Corollary 3.6.1. *If the normal bundle of a soul S of M is flat, then M splits locally isometrically over S; i.e., M is locally isometric to $S \times P^k$ with the product metric, where P^k denotes $\mathbb{R}^k$ together with some metric of nonnegative curvature.*

Proof. By Proposition 3.6.1, the submersion $M \to S$ is flat. The claim now follows from Theorem 2.2.2. □

Example 3.6.1. When S is simply connected, then $\nu(S)$ is trivial whenever it is flat. The converse is not true in general: Consider $S^2 \times \mathbb{R}^2 \times \mathbb{R}$ with the product metric, where each factor has the standard metric. Then $\mathbb{R}$ acts freely by isometries on this product via

$$(t, (p, u, t_0)) \mapsto (\phi_t(p), e^{it}u, t + t_0),$$

with ϕ_t denoting rotation by angle t in S^2 about some axis. The quotient manifold M therefore has nonnegative curvature. It is diffeomorphic to $S^2 \times \mathbb{R}^2$, and has a topological 2-sphere as soul. The normal bundle of the soul is trivial but not flat. In fact, the fibers of $M \to S$ are not even totally geodesic, see [131]. Nonnegatively curved metrics on $S^2 \times \mathbb{R}^2$ were classified in [61], and are all similar to the above example: for such a metric g, there exists a metric g_0 on S^2, a metric g_f on $\mathbb{R}^2$, and a (possibly trivial) isometric action of $\mathbb{R}$ on the Riemannian product $(S^2, g_0) \times (\mathbb{R}^2, g_f)$ such that $(S^2 \times \mathbb{R}^2, g)$ is isometric to $((S^2, g_0) \times (\mathbb{R}^2, g_f) \times \mathbb{R})/\mathbb{R}$ endowed with the submersion metric. As a consequence, if a complete nonnegatively curved metric g on $S^2 \times \mathbb{R}^2$ is not given by a product metric, then there is an isometric effective action of a two-torus on $(S^2 \times \mathbb{R}^2, g)$.

The metric rigidity that is apparent from the above results is nevertheless relative: In a manifold with curvature bounded below by a positive constant, one can deform the metric slightly in a compact set and still retain positive curvature. In a manifold of nonnegative curvature (and in particular one where each point has planes of zero curvature) on the other hand, deforming the metric usually introduces some negative curvature. It is therefore noteworthy that on $M = S^3 \times_{S^1} \mathbb{R}^2$ with the metric from Examples and Remarks 1.8.1 (iii) for which the projection $\rho : S^3 \times \mathbb{R}^2 \to M$ is Riemannian, altering the length of the vector field $\rho_*(0, \partial_\theta)$ by a sufficiently small amount does not change the sign of the curvature, even though the planes that contain the gradient of the distance function to the soul at any point have vanishing curvature (here, ∂_θ is the standard polar coordinate vector field on $\mathbb{R}^2$); for details, see [131].

Theorem 3.6.1 ([71])**.** *If the total space of a vector bundle admits a metric of nonnegative curvature, then so does the total space of its associated sphere bundle.*

Proof. It suffices to show that if M is an open manifold of nonnegative curvature with soul S, then for small enough $r > 0$, the set $S_r := \{p \in M \mid d(p, S) = r\}$ is nonnegatively curved in the induced metric. By the Gauss equations, this will follow once we establish that the second fundamental form l of S_r associated to the inward-pointing unit normal vector field N is positive semi-definite. So choose r small enough that

1. the exponential map of $\nu(S)$ is a diffeomorphism onto $B_r(S)$ when restricted to vectors of length less than r, and
2. metric balls of radius r centered at points in the soul are strongly convex (so that, as mentioned in Section 3.1, their boundary has positive definite second fundamental form).

The Riemannian submersion $\pi : M \to S$ restricts to $S_r \to S$. By Theorem 3.5.1, N is parallel along any horizontal geodesic, so that $l(x, \cdot) = 0$ for any $x \in \mathcal{H}$. The vertical space at $p \in S_r$ coincides with the tangent space at p of $S_r \cap \partial B_r(\pi(p))$. Since $B_r(\pi(p))$ is strongly convex, $l(u, u) > 0$ for any vertical nonzero u, and the second fundamental form is positive semi-definite, as claimed. □

Cheeger and Gromoll asked in [37] whether there is a converse to the soul theorem for spheres: does any vector bundle over a sphere admit a complete metric with nonnegative curvature? At the time of writing, there are no known obstructions to such metrics, and in fact, it has been shown that any vector bundle over a sphere of dimension less than or equal to five admits such a metric, cf. [67]. More generally, one can ask whether any vector bundle over a compact manifold of nonnegative curvature admits a metric with $K \geq 0$. For example, Yang proved that any rank two bundle over the connected sum of complex projective space with itself (but with opposite orientation) admits such a metric [143]. The first counterexamples were provided in [99]. The proof we give here is based on Theorem 3.6.1.

Theorem 3.6.2. *Among the (infinitely many) rank two vector bundles over the torus $S^1 \times S^1$, only the trivial one admits a metric of nonnegative curvature.*

Proof. If ξ is a plane bundle over the torus whose total space admits a metric of nonnegative sectional curvature, then the total space E^1 of the associated unit sphere bundle also admits such a metric by Theorem 3.6.1. By a theorem of Cheeger and Gromoll [37], the universal cover of E^1 splits as the product of a compact simply-connected manifold with Euclidean space. Since E^1 fibers over the torus with fiber S^1, the universal cover must be $\mathbb{R}^3$. Thus, the unit sphere bundle and therefore also the original bundle is trivial. □

For further results along these lines, the reader is referred to [13], [14], and [138]. Up to a finite cover, a soul S splits as $S = C \times T$, where T is a torus, C is a simply-connected nonnegatively curved space, and the normal bundle splits by a base-preserving diffeomorphism as $\xi_C \times T$, where ξ_C is a vector bundle over C with total space $E(\xi_C)$ having nonnegative curvature. In [14], the authors say that a vector bundle virtually comes from C if the pullback bundle under a map $\mathrm{id} \times p \colon C \times T \to C \times T$ (where $p \colon T \to T$ is a finite cover) is $\xi_C \times T$. A typical result is that if ξ is a rank two vector bundle over $C \times T$ whose total space admits a metric with nonnegative curvature, then ξ virtually comes from C. This in turn implies that many vector bundles don't admit nonnegative curvature metrics.

Wilking, on the other hand, has generalized this to Ricci curvature, by showing that any compact manifold with nonnegative Ricci curvature is finitely covered by the product of a torus with a simply-connected space. This product is, in general, topological rather than isometric. Consider for example the twisted product $M = S^2 \times_{\mathbb{Z}} \mathbb{R}$, where the integer $m \in \mathbb{Z}$ acts on S^2 by rotation of angle $2m\alpha\pi$ about some axis and by translation by m on $\mathbb{R}$. If α is irrational, then M is diffeomorphic to $S^2 \times S^1$, but no cover of M is isometric to the Riemannian product $S^2 \times S^1$.

Returning to the original problem posed by Cheeger and Gromoll concerning rank k bundles over n-spheres, it was shown in [72] that if one also imposes an upper bound $\kappa > 0$ for the curvature (of a ball of fixed radius centered at some point of the soul), then there are only finitely many such bundles admitting metrics with $0 \le K < \kappa$, cf. also [121] for similar results involving a non-spherical base. It turns out that the curvature along some vertical planes at each point of the soul must increase as the bundle becomes more twisted.

A different approach to the problem is to try to construct connection metrics of nonnegative curvature on the total space E of the bundle, as was done in Theorem 2.7.3 for the Ricci curvature. In [119], it is shown that a necessary condition for nonnegative curvature is that the curvature tensor R^∇ of the connection satisfy a differential inequality. Specifically, endow the fiber $\mathbb{R}^k$ with the metric given in polar coordinates by $dr^2 + G^2(r)d\sigma^2$, where $d\sigma^2$ is the standard metric on S^{k-1}, and $G^2(r) = \epsilon^2 r^2/(\epsilon^2 + r^2)$. If we define $R^\nabla(u,v) : B_p \to B_p$ by $\langle R^\nabla(u,v)x, y\rangle := \langle R^\nabla(x,y)u, v\rangle$, and denote by k_B the non-normalized sectional curvature of the base B, then R^∇ must satisfy

$$\langle \nabla_x R^\nabla(x,y)u, v\rangle^2 \le |R^\nabla(u,v)x|^2 \left(k_B(x,y) - \frac{3}{4}\epsilon^2 |R^\nabla(x,y)u|^2 \right), \tag{3.6.1}$$

for $u, v \in E_p$, $x, y \in B_p$, $p \in B$. Of course, the simplest way to guarantee this inequality is to require that R^∇ be parallel. This turns out to be fairly restrictive, at least in some cases: If, for example, B is a symmetric space G/H, then $E = G \times_\rho \mathbb{R}^k$ for some orthogonal representation ρ of H, cf. [69].

In [122], it is shown that the above inequality is almost sufficient: If the bundle admits a connection such that

$$\langle \nabla_x R^\nabla(x,y)u, v\rangle^2 < |R^\nabla(u,v)x|^2 \cdot k_B(x,y), \tag{3.6.2}$$

then there exists a connection metric of nonnegative curvature on E. We do not know whether the inequality (3.6.1) is also sufficient. It applies for example to the case where the base is flat (and says that the connection must then be parallel), whereas (3.6.2) can only be used when the base is positively curved. It should finally be noted that the metrics of nonnegative curvature constructed on rank four bundles over S^4 in [67] are not connection metrics.

Chapter 4

Metric Foliations in Space Forms

We have so far focused our attention mostly on the base space B of a Riemannian submersion $M \to B$, in particular when searching for new metrics of nonnegative curvature on B. It is also interesting to look at the total space of the fibration. The very fact that there exists a Riemannian submersion from M (or more generally, that M admits a metric foliation) is a sign that the space possesses a fair amount of symmetry. One therefore expects those Riemannian manifolds with the largest amount of symmetry – namely, space forms – to be the ones that display the most variety as far as these foliations are concerned. Surprisingly, a complete classification of metric foliations on spaces of constant curvature is not yet available. There does, however, exist a classification of metric *fibrations*, at least in nonnegative curvature, which will be described in this chapter.

4.1 Isoparametric foliations

Recall from Section 1.4 that the mean curvature of a metric foliation on M is the one-form κ given by $\kappa(e) = \operatorname{tr} S_{e^{\mathfrak{h}}}$, $e \in TM$. It essentially measures the infinitesimal rate of change of the volume form of the leaves in horizontal directions. To see this, assume the foliation is oriented (which is always the case, at least up to a cover of M). Let ω denote the form on M that is locally the metric dual of $U_1 \wedge \cdots \wedge U_k$, where $U_1, \dots, U_k$ is a local oriented orthonormal basis of the vertical distribution; i.e.,

$$\omega(E_1, \dots, E_k) = \det(\langle U_i, E_j \rangle), \qquad E_j \in \mathfrak{X}M.$$

We denote by $\omega^{\mathfrak{v}}$ the restriction of ω to the vertical distribution. With this notation, we have:

Proposition 4.1.1 (Rummler [112]). *The vertical restriction of the Lie derivative of* ω *in a horizontal direction* $X \in \mathfrak{X}M$ *satisfies*

$$(\mathcal{L}_X\omega)^{\mathrm{v}} = -\kappa(X)\omega^{\mathrm{v}}.$$

Proof. If $U_1, \dots, U_k$ denotes an oriented local orthonormal frame, then

$$\begin{aligned}(\mathcal{L}_X\omega)(U_1,\dots,U_k) &= X\omega(U_1,\dots,U_k) - \sum_{i=1}^{k}\omega(U_1,\dots,[X,U_i]^{\mathrm{v}},\dots,U_k)\\ &= -\sum_{i=1}^{k}\omega(U_1,\dots,[X,U_i]^{\mathrm{v}},\dots,U_k).\end{aligned}$$

Furthermore, we may replace $[X, U_i]^{\mathrm{v}}$ by its projection onto U_i, which equals

$$\langle [X,U_i],U_i\rangle U_i = -\langle \nabla_{U_i}X, U_i\rangle U_i = \langle S_X U_i, U_i\rangle U_i,$$

so that

$$(\mathcal{L}_X\omega)(U_1,\dots,U_k) = -(\operatorname{tr} S_X)\omega(U_1,\dots,U_k),$$

as claimed. □

Definition 4.1.1. A metric foliation is said to be *isoparametric* if its mean curvature form is basic.

By definition, a 1-form κ is basic if its metrically dual vector field is basic; i.e., if $\kappa(X)$ is locally constant along leaves for basic X. In view of Proposition 4.1.1, this amounts to saying that the infinitesimal rate of change of the volume of leaves in basic directions is independent from the point on the leaf at which it is measured. For example, any homogeneous foliation is isoparametric. This follows from Proposition 2.3.4, which actually asserts a stronger property, namely that for basic X and left-invariant U, $S_X U$ is left-invariant. The converse is not true in general: If M is an open manifold of nonnegative curvature with soul S, and if the metric projection $M \to S$ has totally geodesic fibers, then the resulting fibration is isoparametric, but not homogeneous unless M splits locally isometrically as a product over S (one way to see this is to notice that if $M \to S$ is homogeneous, then by Proposition 2.3.4, $A_X Y$ is left-invariant for basic X and Y, and in particular has constant norm along geodesics emanating from the soul; it must then be identically zero by Proposition 3.6.1. Now apply Theorem 2.2.2). A typical example is $TS^n = SO(n+1) \times_{SO(n)} \mathbb{R}^n \to S^n$. When the foliation is one-dimensional, however, the converse is true under weak curvature restrictions, see also [134], [55]:

Proposition 4.1.2. *Any one-dimensional isoparametric Riemannian foliation* $\mathcal{F}$ *with complete leaves on a manifold with curvature bounded below is locally homogeneous; i.e.,* $\mathcal{F}$ *is locally generated by a Killing field.*

Proof. Since the mean curvature form κ is basic, so is $d\kappa$, and by Proposition 1.4.1, the function

$$\operatorname{div} A_X Y = -\frac{1}{2} d\kappa(X,Y)$$

is then constant along leaves for basic X and Y. If T is a local unit vertical field, this translates into $\langle \nabla_T A_X Y, T\rangle$ being constant, since $\langle \nabla_Z A_X Y, Z\rangle$ is always zero for horizontal Z. Thus, if $c : \mathbb{R} \to M$ is a unit-speed curve parametrizing a complete leaf, then $\langle A_X Y \circ c, \dot{c}\rangle'$ is constant, and $\langle A_X Y \circ c, \dot{c}\rangle$ is an affine function. It must then be constant by O'Neill's formula, if the curvature is bounded below. Consequently, $d\kappa(X,Y) = 0$, and κ is closed, because

$$d\kappa(X,T) = X\kappa(T) - T\kappa(X) - \kappa[X,T] = -T\kappa(X) = 0.$$

Thus, $\kappa = d\phi$ locally, for a function ϕ that is constant along leaves. Set $L := e^{-\phi}$, $U := LT$. Then U is Killing, since

$$\langle \nabla_X U, X\rangle = L\langle \nabla_X T, X\rangle = L\langle \nabla_T X, X\rangle = 0,$$

and

$$\langle \nabla_X U, T\rangle + \langle \nabla_T U, X\rangle = XL\langle T,T\rangle - \langle U, \nabla_T X\rangle = XL + L\kappa(X) = 0. \qquad \square$$

The relevance of isoparametric foliations for space forms is illustrated by the following:

Theorem 4.1.1 ([63]). *Any metric foliation of a space form of nonnegative curvature is isoparametric.*

Proof. Let x be horizontal, and consider a Riemannian submersion that locally defines the foliation in a neighborhood of the footpoint of x. We will prove a stronger assertion, namely that if λ is an eigenvalue of S_x, then it is also an eigenvalue (of equal multiplicity) of $S_{\tilde{x}}$, for any horizontal $\tilde{x}$ with $\pi_*\tilde{x} = \pi_* x$. Denote by γ (resp. $\tilde{\gamma}$) the geodesic with initial tangent vector x (resp. $\tilde{x}$), and consider the Jacobi field J along γ with $J(0) = u$, $J'(0) = -S_x u = -\lambda u$, where u denotes a unit λ-eigenvector of S_x. Notice that $J = \phi E$, where E is the parallel field along γ with $E(0) = u$, and ϕ is the solution of the O.D.E.

$$\phi'' + c\phi = 0, \qquad \phi(0) = 1, \quad \phi'(0) = -\lambda,$$

with c denoting the curvature of the space. Assume for now that $\lambda \neq 0$ if $c = 0$. Then $J(l) = 0$ for some $l \in \mathbb{R}$. But J is by definition projectable, so that $\pi_* J$ is Jacobi along $\pi \circ \gamma$ by Theorem 1.6.1. By Lemma 1.6.1, there exists a unique Jacobi field $\tilde{J}$ along $\tilde{\gamma}$ with $\pi_*\tilde{J} = \pi_* J$ and $\tilde{J}(l) = 0$. In particular, $\tilde{J}(0)$ must be vertical (because J is), so that $\tilde{J}'^{\mathbf{v}}(0) = -S_{\tilde{x}}\tilde{J}(0)$. This, together with the fact that $\tilde{J}(l) = 0$ implies that $\tilde{J} = \phi\tilde{E}$ for some parallel field E. It follows that $\tilde{J}(0)$ is a λ-eigenvector of $S_{\tilde{x}}$. Let k denote the multiplicity of λ as an eigenvalue of S_x. Since for any Jacobi field Y along $\pi \circ \gamma$ that vanishes at 0 and l there exists a unique projectable Jacobi field J along γ with $J(0)$ in the λ-eigenspace of S_x, l is a conjugate point of $\pi \circ \gamma$ of order $\leq k$. Conversely, if J is a projectable Jacobi field along γ with $J'(0) = -S_x J(0)$ and $J(0) \neq 0$, then $\pi_* J$ is a nontrivial Jacobi field that vanishes at 0 and l (if $\pi_* J \equiv 0$, then J is vertical; i.e., J is a holonomy field, and since $J(l) = 0$, $J \equiv 0$). Thus, l is a conjugate point of order k. Applying Lemma 1.6.1 again, we see that the multiplicity of λ as an eigenvalue of $S_{\tilde{x}}$ is also k. This establishes the theorem in all cases except perhaps when $c = \lambda = 0$. But then it must also be true in the latter case. □

Corollary 4.1.1. *A one-dimensional metric foliation of a space form of nonnegative curvature is locally homogeneous.*

Proof. This follows immediately from Theorem 4.1.1 and Proposition 4.1.2. Notice that if the space is simply connected, then the Killing field is globally defined. □

Even though the above is not necessarily true in constant negative curvature (see Examples and Remarks (i) below), a slightly more general result does hold, cf. also [55]:

Theorem 4.1.2. *A one-dimensional metric foliation of a hyperbolic space form is either locally homogeneous or flat.*

Proof. We first claim that in constant (not necessarily negative) curvature c, the A-tensor vanishes everywhere as soon as it vanishes at any one point p. To see this, it is enough to show that it is zero along any horizontal geodesic γ emanating from p, since for basic X, Y, $A_X Y$ has constant norm along leaves by O'Neill's formula. An equivalent claim is that the totally geodesic hypersurface $\exp_p(\mathcal{H}_p)$ is horizontal everywhere, or alternatively, that any parallel vector field E along γ with $E(0)$ horizontal remains horizontal for all t. But if J is a holonomy Jacobi field along γ, then

$$\langle J, E\rangle'' = \langle J'', E\rangle = -\langle R(J, \dot{\gamma})\dot{\gamma}, E\rangle = -c\langle J, E\rangle.$$

The claim now follows from the initial conditions, because $\langle J, E\rangle(0) = 0$ by assumption, and

$$\langle J, E\rangle'(0) = \langle J', E\rangle(0) = -\langle (S_{\dot{\gamma}} + A^*_{\dot{\gamma}})J, E\rangle(0) = -\langle A^*_{\dot{\gamma}} J, E\rangle(0) = 0$$

if $A_p \equiv 0$.

Resuming the proof of the theorem, assume that the foliation is not flat. By the above claim, there exist at any point p basic vector fields X, Y with $A_X Y \neq 0$ at p. Theorem 1.5.1 then implies that

$$R^{\mathrm{v}}(X, Y)X = -(\nabla^{\mathrm{v}}_X A)_X Y + 2S_X A_X Y,$$

so that if T is a local unit field spanning the vertical distribution, then

$$0 = \langle R(X, Y)X, T\rangle = -\langle (\nabla^{\mathrm{v}}_X A)_X Y, T\rangle + 2\langle S_X A_X Y, T\rangle. \tag{4.1.1}$$

Now, the first term on the right in (4.1.1) is locally constant along leaves, since it can be written as

$$\begin{aligned}\langle(\nabla^{\mathbf{v}}_X A)_X Y, T\rangle &= \langle\nabla_X(A_X Y), T\rangle - \langle A_{\nabla_X X} Y, T\rangle - \langle A_X \nabla_X Y, T\rangle \\ &= X\langle A_X Y, T\rangle - \langle A_{\nabla_X X} Y, T\rangle - \langle A_X \nabla_X Y, T\rangle,\end{aligned}$$

where both $\nabla^{\mathbf{h}}_X X$ and $\nabla^{\mathbf{h}}_X Y$ are basic, whereas

$$TX\langle A_X Y, T\rangle = [T, X]\langle A_X Y, T\rangle + XT\langle A_X Y, T\rangle = 0$$

because $[T, X]$ is vertical. Thus, $\langle S_X A_X Y, T\rangle$ is constant along leaves by (4.1.1), so that $\kappa(X)$ is also constant, since

$$\langle S_X A_X Y, T\rangle = \langle A_X Y, S_X T\rangle = \langle A_X Y, T\rangle\kappa(X).$$

Summarizing, $\kappa(X)$ is locally constant for all X in a nonempty open subset of basic fields along the leaf through p, and κ is therefore basic. □

Examples and Remarks 4.1.1. (i) In a space of constant curvature c, any submanifold with flat normal bundle is locally a leaf of a (flat) metric foliation, as remarked in Examples 2.2.1(ii). This foliation cannot be extended to the whole space if $c > 0$ by Theorem 2.2.2. For $c = 0$, it can iff L is totally geodesic. When $c < 0$, however, there is no such rigidity. This also shows that Theorem 4.1.1 does not hold in this case: Consider for example a line in hyperbolic space. Deform the line in a neighborhood of some point so that it is no longer totally geodesic there, but do it slightly so that the exponential map of the normal bundle is still one-to-one. Exponentiating parallel sections of the normal bundle then yields a metric foliation of hyperbolic space that is not isoparametric.

(ii) The Hopf fibrations with fibers S^1 and S^3 are homogeneous. Even though the higher-dimensional Hopf fibration $S^{15} \to S^8$ with fiber S^7 is isoparametric (in fact, it is totally geodesic) by Theorem 4.1.1, it is not homogeneous. Before arguing this, notice first that since the fibration is a fat bundle, it is *weakly substantial*; i.e., the image of the A-tensor equals the whole vertical distribution. This implies that it cannot be homogeneous. In fact, we claim that if a homogeneous submersion is weakly substantial, then the fiber is a Lie group (the fiber of the Hopf fibration is S^7, which of course is not a Lie group). To see this, let G be the group of isometries generating the fibration, so that a fiber has the form G/H, where H is the isotropy group at some point p. Consider $h \in H$. Since $\pi \circ h = \pi$ and since h preserves the vertical, and hence horizontal distributions, it must preserve basic fields; i.e., $h_* X = X \circ h$ for any basic field X (and in particular, h_* is the identity on the horizontal space at p). h_* must then also preserve the bracket of basic fields, so that $h_* A_x y = A_x y$ for any horizontal vectors x, y at p. Thus, h_* is the identity on the vertical space also, and since h is an isometry, it is the identity map. This means that H is trivial, and the fiber is G, as claimed.

(iii) Recall that given a metric foliation on M, a one-form α is basic if its metrically dual vector field $\alpha^\sharp$ is basic. This is easily seen to be equivalent to requiring that $\alpha(T) = 0$ and $d\alpha(T, E) = 0$ for any vertical field T and arbitrary field E. More generally, a differential form α on M is said to be *basic* if

$$\imath_T \alpha = 0, \quad \imath_T d\alpha = 0, \qquad \text{for vertical } T,$$

($\imath$ denotes interior multiplication). By definition, the differential of a basic form is again basic, so that d, when restricted to the algebra of basic forms, induces a so-called *basic cohomology* of the leaf space introduced by Reinhart [110]. A number of authors have studied this complex in an attempt to develop a basic Hodge theory and basic Laplacian, leading to a representation of basic cohomology classes by harmonic forms, see, e.g., [79] in the isoparametric case and [101] in the general case.

4.2 Metric fibrations of Euclidean space

Our next objective is to derive a classification of Riemannian submersions $\pi : \mathbb{R}^{n+k} \to B^n$. A simple, yet illustrative example to keep in mind throughout this discussion is the orbit fibration $\pi : \mathbb{R}^3 \to B^2 = \mathbb{R}^3/\mathbb{R}$, where $\mathbb{R}$ is the Lie group of isometries acting on $\mathbb{R}^3 = \mathbb{C} \times \mathbb{R}$ via glide-rotations; i.e.,

$$\begin{aligned} \mathbb{R} \times (\mathbb{C} \times \mathbb{R}) &\longrightarrow \mathbb{C} \times \mathbb{R}, \\ (t, (x, t_0)) &\longmapsto (e^{it}x, t_0 + t), \end{aligned}$$

cf. Examples and Remarks 1.5.1(iv). Notice that there is exactly one totally geodesic fiber, namely the z-axis. It turns out it is the fiber over the soul of the nonnegatively curved manifold B^2.

In general, if $\pi : \mathbb{R}^{n+k} \to B^n$ is a Riemannian submersion, then B^n has nonnegative curvature, and π factors as a fibration over the universal cover of B, followed by a covering map. Covering maps in nonnegative curvature are well understood (see, e.g., [37]), and we may therefore assume without loss of generality that B is simply connected. It follows from the long exact homotopy sequence of π that the fiber of the submersion is connected. Using the spectral sequence for the homology of the fibration, one concludes that B is contractible, cf. [70]. Since B is also a vector bundle over a soul, it must be diffeomorphic to Euclidean space, and any soul consists of a point.

Proposition 4.2.1. *If $\pi : \mathbb{R}^{n+k} \to B$ is a Riemannian submersion, then the fiber over any soul of B is an affine subspace.*

Proof. The general idea is to lift the soul construction to Euclidean space, cf. also [38]: Let $\{p\}$ denote a soul of B, $c : [0, \infty) \to B$ a ray emanating from p. Notice that any lift $\tilde{c}$ of c must also be a ray: this is of course trivial in this case, since any normal geodesic in Euclidean space is a ray, but is also true in general: otherwise, for some t, the line segment from $\tilde{c}(0)$ to $\tilde{c}(t)$ would be shorter than t, implying that $c = \pi(\tilde{c})$ is not minimal. If

$$B_c = \bigcup_{t>0} B_t(c(t))$$

is the open half-space determined by c from the soul construction, then $\pi(B_{\tilde{c}}) = B_c$, since π maps metric balls onto metric balls of the same radius. Denote by $\tilde{B}_c$ the union of all $B_{\tilde{c}}$, where $\tilde{c}$ ranges over all lifts of c, and by $\tilde{C}_c$ its complement in $\mathbb{R}^{n+k}$. $\tilde{C}_c$ is closed and convex (as an intersection of closed half-spaces), and by construction, $C_c \subset \pi(\tilde{C}_c)$, where C_c is the complement of B_c in B. The reverse inclusion also holds, for otherwise, one could find some $q \in \tilde{C}_c$ such that $\pi(q) \in B_c$; i.e., there would exist some t_0 such that $d(\pi(q), c(t_0)) < t_0$. But then the horizontal lift to q of a minimal geodesic from $\pi(q)$ to $c(t_0)$ is a curve of length less than t_0 connecting q to $\tilde{c}(t_0)$ for some lift $\tilde{c}$ of c, contradicting the fact that $q \in C_{\tilde{c}}$. Next, set

$$\tilde{C} := \bigcap_c \tilde{C}_c, \qquad C := \bigcap_c C_c,$$

where c ranges over all rays emanating from p. Just as above, one has that $\tilde{C}$ is a closed convex set of Euclidean space with $\tilde{C} = \pi^{-1}(C)$ and $\partial\tilde{C} = \pi^{-1}(\partial C)$. If $C = \{p\}$, then $\tilde{C} = \pi^{-1}(p)$ is a closed, connected, convex submanifold without boundary of Euclidean space, i.e., an affine subspace. If C has nonempty boundary, define sets

$$\tilde{C}^a = \{q \in \tilde{C} \mid d(q, \partial\tilde{C}) \geq a\}, \qquad C^a = \{q \in C \mid d(q, \partial C) \geq a\},$$

where $0 \leq a \leq \max\{d(q, \partial C) \mid q \in C\}$. Both sets are closed and totally convex by the results from Chapter 3. Furthermore, given any two points p_0, p_1 in B, the distance between them equals the distance between the fibers over them, as well as the distance between any point on one fiber and the other fiber. This is easily seen to imply that $\tilde{C}^a = \pi^{-1}(C^a)$. Iterating this procedure finitely many times until the set in the base no longer has boundary (and therefore equals $\{p\}$) lets us draw the same conclusion as when C consists of the single point p, thereby establishing the result. □

The above proposition allows us to strengthen Theorem 4.1.1 in the case of a fibration of Euclidean space:

Proposition 4.2.2. *The mean curvature form κ of a Riemannian submersion $\pi : \mathbb{R}^{n+k} \to B^n$ is basic and exact; i.e., there exists a function $f : B \to \mathbb{R}$ such that $\kappa = d(f \circ \pi)$.*

Proof. The fact that κ is basic follows from Theorem 4.1.1, so we only have to show that it is closed. Since κ vanishes when applied to vertical vectors, $d\kappa(U, V) = 0$ for vertical U, V. Furthermore, for basic X, the bracket $[X, U]$ is vertical, so that

$$d\kappa(X, U) = X\kappa(U) - U\kappa(X) = -U\kappa(X) = 0,$$

because κ is basic. It remains to show that $d\kappa(X,Y) = 0$ for basic X, Y, or equivalently (by Proposition 1.4.1), that the vertical field A_XY has vanishing divergence. Now, this divergence is the one induced by the fiber metric, since $\langle \nabla_Z A_X Y, Z\rangle = -\langle A_XY, \nabla_Z Z\rangle = 0$ for basic Z. Furthermore, the divergence is constant along fibers because κ, and hence also $d\kappa$, is basic. Denote by F the totally geodesic fiber over the soul, and consider a minimal segment c from F to some fiber L at distance l from F. The horizontal lifts of $\pi \circ c$ induce a holonomy diffeomorphism $h : F \to L$, and by Lemma 1.4.2, the derivative of h at any point q of F satisfies $h_*u = J(l)$, where J is the holonomy field along the line $t \mapsto \exp(tZ_q)$, with $J(0) = u$ (here, Z denotes the basic field along F that extends $\dot{c}(0)$). Now, F is totally geodesic, so that $J'(0) = -A_Z^*u$, and $J(t) = E + tF$, where E and F are the parallel fields with $E(0) = u$ and $F(0) = -A_Z^*u$. In particular E and F are mutually orthogonal, so that

$$|h_*u|^2 = |u|^2 + l^2|A_Z^*u|^2.$$

Thus, the norm of h_* is bounded below by 1, and since $|A_XY|$ is constant along fibers, it is also bounded above by some constant. It follows that if $B_r \subset L$ denotes the h-image of a ball of radius r in F about some point, then $\operatorname{vol} B_r \geq a \cdot r^k$ and $\operatorname{vol} \partial B_r \leq b \cdot r^{k-1}$ for some positive constants a and b. If N_r denotes the outward unit normal field to ∂B_r, then Stokes' Theorem implies

$$a \cdot |\operatorname{div} A_XY| \cdot r^k \leq \left| \int_{B_r} \operatorname{div} A_XY \right| = \left| \int_{\partial B_r} \langle A_XY, N_r\rangle \right| \leq b \cdot |A_XY| \cdot r^{k-1},$$

so that $\operatorname{div} A_XY \equiv 0$ if the above inequality is to hold for all $r > 0$. □

Up to a congruence of $\mathbb{R}^{n+k}$, the totally geodesic fiber F is $\{0\} \times \mathbb{R}^k \subset \mathbb{R}^{n+k}$. Normalize the function $f : B^n \to \mathbb{R}$ from Proposition 4.2.2 so that it equals zero at $\pi(F)$. If ω is the vertical volume form from Proposition 4.1.1, define the *holonomy form* of the fibration to be the k-form η given by

$$\eta := e^{-(f\circ\pi)}\omega.$$

It can alternatively be described as follows: Consider an oriented orthonormal parallel basis $E_1, \dots, E_k$ of vector fields along F, and extend them radially via holonomy diffeomorphisms from F; i.e., define vector fields U_i on $\mathbb{R}^{n+k}$ by

$$U_i(x,u) := ||\big(E_i(0,u) - A^*_{\mathcal{J}_{(0,u)}x} E_i(0,u)\big), \qquad i = 1, \dots, k, \tag{4.2.1}$$

where $||$ denotes parallel translation from $(0,u)$ to (x,u), and $\mathcal{J}_{(0,u)}$ is the canonical isomorphism of Euclidean space with its tangent space at $(0,u)$. Thus, for a line c emanating orthogonally from F, $U_i \circ c$ is the holonomy Jacobi field that equals E_i at 0. The relation between the U_i and η is given by the following:

Lemma 4.2.1. $\eta^\sharp = U_1 \wedge \cdots \wedge U_k$.

Proof. Since η and the dual of $U_1 \wedge \cdots \wedge U_k$ are both vertical forms, it suffices to show that at any point p,

$$\eta(v_1, \dots, v_k) = \langle U_1(p) \wedge \cdots \wedge U_k(p), v_1 \wedge \cdots \wedge v_k \rangle,$$

where $v_1, \dots, v_k$ denotes a positively oriented orthonormal basis of the fiber at p; equivalently, that $e^{-(f\circ\pi)} = \omega(U_1, \dots, U_k)$. Now by definition, both functions are constant equal to 1 along F. Furthermore, if X is the tangent field of a horizontal geodesic emanating from F, then

$$X\big(e^{-(f\circ\pi)}\big) = -e^{-(f\circ\pi)} X(f \circ \pi) = -e^{-(f\circ\pi)} \kappa(X),$$

whereas by Proposition 4.1.1,

$$\begin{aligned} X\big(\omega(U_1, \dots, U_k)\big) &= \mathcal{L}_X\big(\omega(U_1, \dots, U_k)\big) = (\mathcal{L}_X \omega)(U_1, \dots, U_k) \\ &= -\omega(U_1, \dots, U_k)\kappa(X), \end{aligned}$$

using the fact that $[X, U_i] = 0$. The claim clearly follows. □

Lemma 4.2.1 says that the k-form $U_1 \wedge \cdots \wedge U_k$ is holonomy-invariant in the sense that the wedge product of holonomy fields is independent of the chosen horizontal path. We will soon see that the vector fields U_i are in fact global Killing fields that generate the isometric action. In the special case of a one-dimensional fibration, it is easy to see that U is a Killing field, i.e., that the assignment $z \mapsto \nabla_z U$ is skew-adjoint: If $T = U/|U|$ is the unit field in direction U, then $U = e^{-(f\circ\pi)}T$, so that, for horizontal X,

$$\langle \nabla_X U, X \rangle = -\langle \nabla_X X, U \rangle = 0,$$

whereas

$$\langle \nabla_U U, U \rangle = \frac{1}{2} U\big(e^{-2(f\circ\pi)}\big) = 0$$

since $\pi_* U = 0$. Finally,

$$\langle \nabla_X U, T \rangle + \langle \nabla_T U, X \rangle = X\big(e^{-(f\circ\pi)}\big) + e^{-(f\circ\pi)}\kappa(X) = 0$$

by Lemma 4.2.1. Thus, U is Killing.

For simplicity of notation, we will, for the remainder of the section, identify Euclidean space with its tangent space at any point. Thus, the vector field U_i from (4.2.1) becomes a map from $\mathbb{R}^{n+k}$ to itself, and the holonomy form (or rather its dual $\eta^\sharp$) is a map from $\mathbb{R}^{n+k}$ to $\Lambda_k(\mathbb{R}^{n+k})$. We say a map from $\mathbb{R}^{n+k}$ to a vector space is *polynomial of degree at most r* if each component function $\phi : \mathbb{R}^{n+k} \to \mathbb{R}$ of this map in some basis is a polynomial of degree at most r in the usual sense. For example, each vector field U_i is polynomial of degree at most 1 along any affine subspace $\mathbb{R}^n \times \{u\}$, since the map $x \mapsto A_x^* E$ is linear. It follows that $\eta^\sharp$ is polynomial of degree at most k along these subspaces. In fact, it is polynomial along *any* horizontal space, not just those based at F; this will be a key point in the forthcoming classification of metric fibrations on Euclidean space:

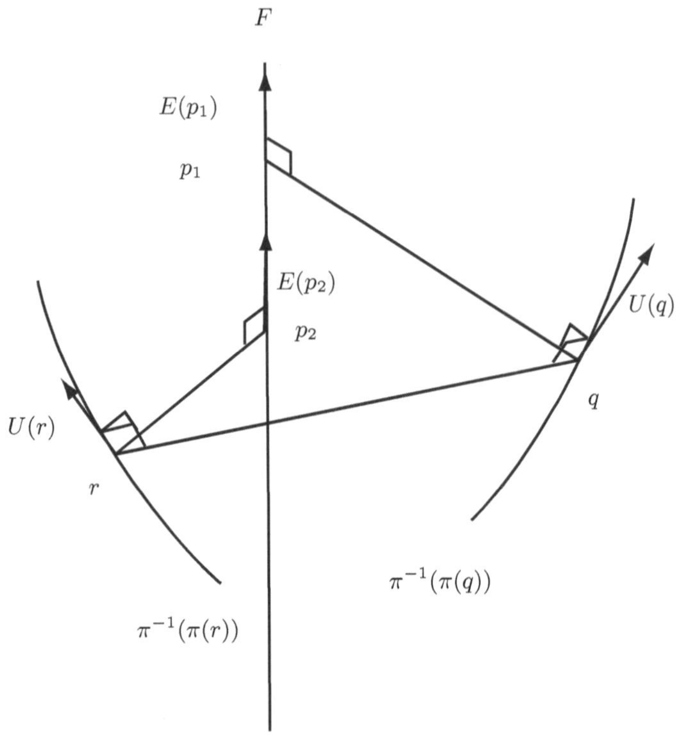

Figure 4.1: Holonomy invariance in dimension one.

Lemma 4.2.2. *$\eta^\sharp$ is polynomial of degree at most k on every affine horizontal subspace.*

Proof. Consider $p \in \mathbb{R}^{n+k}$, and a point q on the horizontal subspace H through p. By Lemma 4.2.1, $\eta^\sharp$ is holonomy-invariant, so that

$$\eta^\sharp(q) = \bigwedge_i U_i(q) = \bigwedge_i \left(U_i(p) - (A^*_{q-p} + S_{q-p})U_i(p)\right].$$

Thus, after translating the origin to p, it suffices to show that the map

$$x \longmapsto \bigwedge_i (E_i - A^*_x E_i - S_x E_i)$$

is polynomial of degree at most k. This in turn follows from the fact that $x \mapsto A^*_x E + S_x E$ is linear. □

Lemma 4.2.3. *$\eta^\sharp$ is polynomial along every affine plane through a point $(0, a) \in F$ spanned by a horizontal x and a vertical u in the image of A_x.*

Proof. Normalize x to have length 1, and denote by $H(t)$ the horizontal space at (tx, a) for $t \geq 0$; i.e., at distance t from F along the (horizontal) line in direction x through $(a, 0)$. By Lemma 4.2.2, $\eta^\sharp$ is polynomial on $H(t)$. Now, x is in $H(t)$ for all t. By a continuity argument, the claim follows once we establish that u is in $H(\infty)$, where the latter denotes the limit of $H(t)$ as $t \to \infty$. In fact, $H(\infty)$ is the direct sum of the kernel of A_x and the image of A_x, because of the form of holonomy fields in Euclidean space: the holonomy field J that equals E at time 0 is $J(t) = E - tA_x^* E$ (where E is extended to be parallel). Now, let t go to infinity, to conclude that the vertical space at infinity is spanned by $(\ker + \operatorname{Im}) A_x^*$. Equivalently, the horizontal space at infinity is spanned by $(\ker + \operatorname{Im}) A_x$. $\square$

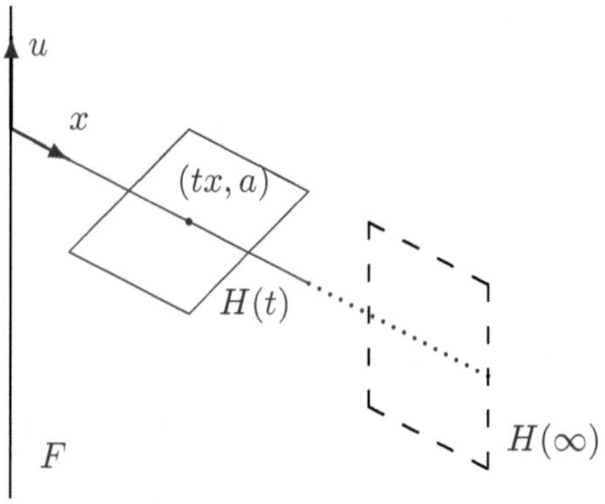

Since F is totally geodesic, any vector x in its normal bundle ν may be extended by parallel translation to a horizontal vector field along F. Such a field will be denoted by the same lowercase letter x to distinguish it from the uppercase notation X for basic fields. Thus, the former are the parallel sections for the usual Euclidean connection $\nabla^{\mathbf{h}}$ on ν, whereas the latter represent those that are parallel for the Bott connection ∇^B from (1.3.3). The connection difference form $\Omega = \nabla^{\mathbf{h}} - \nabla^B$ is then the 1-form on F with values in the bundle of skew-symmetric endomorphisms of ν given by

$$\Omega(u)x = -A_x^* u, \qquad u \in TF, \quad x \in \nu.$$

At this point, it is convenient to simplify matters by getting rid of the "translational" part of the submersion, which is *grosso modo* the kernel of A^*: for a point p in the fiber F, denote by $\mathcal{A}_p$ the (affine) space at p spanned by all integrability fields $A_x y$. Define the kernel of A^* to be the union over $p \in F$ of $\mathcal{A}_p^\perp$. We then have the following

Proposition 4.2.3. *π factors as an orthogonal projection $\mathbb{R}^{n+k-l} \times \mathbb{R}^l \to \mathbb{R}^{n+k-l} \times \{0\}$ followed by a Riemannian submersion $\pi' : \mathbb{R}^{n+k-l} \to B$, where the fiber F' of π' over the soul of B is spanned by the image of A; i.e., for any $p \in F'$, $F' = \mathcal{A}_p$. Furthermore, given parallel sections x, y of the normal bundle ν' of F', $A_x y$ is a parallel vector field along F'.*

Proof. Let x be a vector in the normal bundle ν of F at $(0,a)$, and u a vector in the image of A_x. By Lemma 4.2.3, the holonomy form is polynomial along the plane through $(0,a)$ spanned by x and u, and therefore so is its derivative in direction x. The restriction of this derivative to the line $t \mapsto \gamma_u(t) := (0, a+tu)$ is given by

$$\nabla_x \eta^\sharp = -\sum_i E_1 \wedge \cdots \wedge A_x^* E_i \wedge \cdots \wedge E_k.$$

Now, the E_j are parallel, and $A_x^* E_i$ is horizontal and bounded in norm. Since a bounded polynomial is constant, we conclude that each $A_x^* E_i$ is parallel along γ_u, or equivalently,

$$(A_x y \circ \gamma_u)' \equiv 0, \qquad u \in \operatorname{Im} A_x. \tag{4.2.2}$$

Thus, the image of A_x, though *a priori* not of constant rank along F, is totally geodesic, and consists of a disjoint union of affine subspaces. Next, let $u \in \ker A_x^*$. We claim that $\dot\gamma_u(t) \in \ker A_x^*$ for all t. To see this, consider the variation $V(t,s) = \exp_{su} tx$, which projects down to a variation $W = \pi \circ V$ on the quotient. The Jacobi field $Y(t) = W_* D_2(t,0)$ induced by W satisfies $Y(0) = 0$, and

$$Y'(0) = \pi_* \nabla_{D_1(0,0)} (V_* D_2)^{\mathbf{h}} = -\pi_* \nabla^{\mathbf{h}}_{D_1(0,0)} (V_* D_2)^{\mathbf{v}} = \pi_* A_x^* u = 0.$$

Thus, Y is identically zero, or equivalently, the parallel field x is actually basic along γ_u, so that $A_x^* \dot\gamma_u = -(x \circ \gamma_u)' \equiv 0$. This establishes the claim. The latter in turn implies that the image of A has constant rank: in fact, it says that for any point p in F, $\mathcal{A}_p^\perp = \ker A_p^*$ is totally geodesic since it is the intersection over all unit horizontal x at p of the kernel of A_x^*. Now, up to congruence, $\mathcal{A}_0$ is $\mathbb{R}^{k-l} \times \{0\}$ for some integer l by (4.2.2). It follows that for any $(a,b) \in \mathbb{R}^l \times \mathbb{R}^{k-l} = F$, $\mathcal{A}_{(a,b)}^\perp = \ker A_{(a,b)}^* = \{a\} \times \mathbb{R}^l$, since $\mathcal{A}_{(a,0)}^\perp = \{a\} \times \mathbb{R}^l$. Thus, $\mathcal{A}_{(a,b)} = \mathbb{R}^{k-l} \times \{b\}$, and F splits isometrically as $\mathbb{R}^{k-l} \times \mathbb{R}^l$ with the image of A tangent to the first factor, and the kernel of A^* tangent to the second one. This splitting extends to all of Euclidean space, since the kernel of A^* is invariant under parallel translation along horizontal lines γ that intersect F, thereby establishing the first part of the proposition. After factoring out an orthogonal projection, we now have a submersion $\pi' : \mathbb{R}^{n+k-l} \to B$ where the fiber F' over the soul of B is spanned by the image of the A-tensor. An argument similar to the one that led to (4.2.2) then implies that each integrability field is parallel along any line in F', thereby concluding the proof. □

We are now in a position to classify metric fibrations of Euclidean spaces:

Theorem 4.2.1. *Let $\pi : \mathbb{R}^{n+k} \to B^n$ be a Riemannian submersion with connected fibers. Then there exists an orthogonal representation $\phi : \mathbb{R}^k \to SO(n)$, such that, up to congruence, π is the orbit fibration of the free isometric group action ψ of $\mathbb{R}^k$ on $\mathbb{R}^{n+k} = \mathbb{R}^n \times \mathbb{R}^k$ given by*

$$\psi(v)(x,u) = (\phi(v)x, u+v), \qquad u, v \in \mathbb{R}^k, \quad x \in \mathbb{R}^n.$$

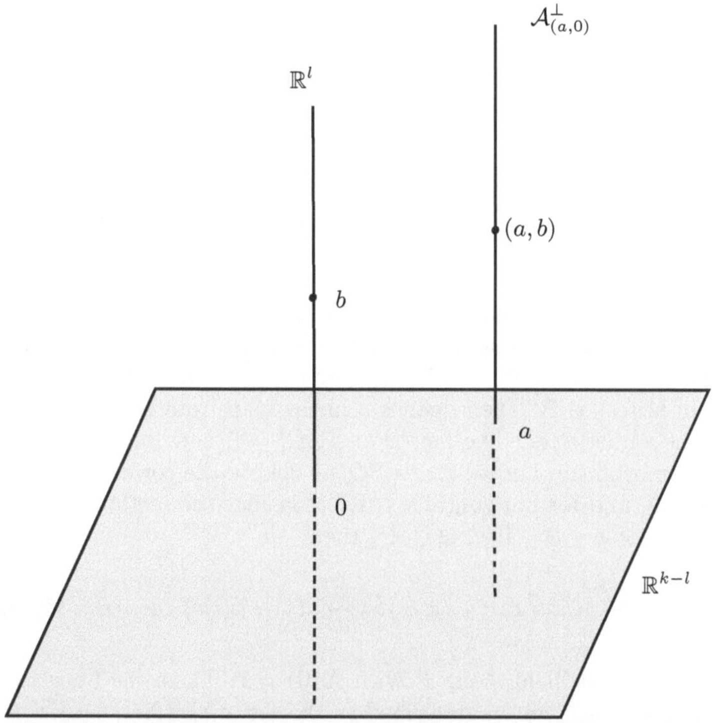

Conversely, of course, given any homomorphism $\phi : \mathbb{R}^k \to SO(n)$, the orbits of the free isometric action ψ described above form a metric fibration of $\mathbb{R}^{n+k}$. Before going into the proof of the theorem, it may be useful to give a rough description of the main idea involved: If we identify the trivial rank n normal bundle ν of F with $\mathbb{R}^n$ by means of parallel translation, then the bundle of skew-adjoint endomorphisms of ν is simply $\mathfrak{so}(n)$. Similarly, TF is identifiable with F via parallel translation. The connection difference form $\Omega = \nabla^{\mathbf{h}} - \nabla^B$ can then be viewed as a linear map $\Omega : F = \mathbb{R}^k \to \mathfrak{so}(n)$. Proposition 4.2.3 now implies that Ω is a Lie algebra homomorphism. The corresponding Lie group homomorphism turns out to be the representation ϕ in the theorem.

Proof. In general, it is a standard fact that if ∇^1 and ∇^2 are connections on a vector bundle with connection difference 1-form $\Omega = \nabla^1 - \nabla^2$, then the curvature tensors of these connections satisfy

$$R^1 - R^2 = d_{\nabla^2}\Omega + [\Omega, \Omega], \tag{4.2.3}$$

where d_∇ denotes the exterior derivative operator associated to ∇; i.e.,

$$d_\nabla\Omega(U, V) = \nabla_U\Omega(V) - \nabla_V\Omega(U) - \Omega[U, V], \tag{4.2.4}$$

cf. [106]. Now, both the Bott and the Euclidean connections on ν are flat (since they admit globally parallel sections), so that if $\Omega = \nabla^{\mathrm{h}} - \nabla^B$, then (4.2.3) becomes

$$d_{\nabla^{\mathrm{h}}}\Omega = -d_{\nabla^B}\Omega = [\Omega, \Omega]. \tag{4.2.5}$$

If U, V are parallel vector fields on F, and x is a parallel section of ν, then by Proposition 4.2.3, $A^*_x V$ is a parallel section of ν, so that

$$(\nabla_U \Omega(V))x = \nabla_U(\Omega(V)x) = -\nabla_U(A^*_x V) = 0.$$

(4.2.4) then implies that $d_{\nabla^{\mathrm{h}}}\Omega = 0$, and (4.2.5) that $[\Omega, \Omega] = 0$. $F = \mathbb{R}^k$ will be identified with its tangent space at any point via parallel translation, and similarly, sections of the normal bundle of F will be viewed as maps from $\mathbb{R}^k$ to $\mathbb{R}^n$. The restriction of Ω to $0 \in \mathbb{R}^k$ then defines a linear map from $\mathbb{R}^k$ to $\mathfrak{so}(n)$, which we denote by the same letter. The fact that $[\Omega, \Omega] = 0$ now implies that it is a Lie algebra homomorphism. Let $\phi : \mathbb{R}^k \to SO(n)$ denote the corresponding Lie group homomorphism, and for horizontal $x \in \mathbb{R}^n$, consider the section X of ν given by $X(u) = \phi(u)x$, for $u \in \mathbb{R}^k$. If v, $w \in \mathbb{R}^k$, then

$$(\nabla_w X)(v) = \frac{d}{dt}_{|0}(t \mapsto \phi(v + tw)x) = \frac{d}{dt}_{|0}(t \mapsto \phi(tw) \cdot \phi(v)x = \Omega(w)X(v),$$

so that X is the basic field along F with $X(0) = x$. Thus, the fiber through any point $(x, u) \in \mathbb{R}^n \times \mathbb{R}^k$ can be described as the set of all $\big(X(u + v), u + v\big)$ where X is the basic field with $X(u) = x$ and v ranges over $\mathbb{R}^k$. This completes the proof of the theorem, since the free action ψ in the statement satisfies

$$\psi(v)(x, u) = (\phi(v)x, u + v) = (\phi(u + v)\phi(-u)x, u + v) = (X(u + v), u + v).$$

Here, we have used the fact that $X(0) = \phi(-u)x$, which follows from $X(u) = \phi(u)X(0) = x$. □

It was already observed in Example 2.3.1 that along any given fiber of π, there exists a point-wise orthonormal basis of Killing fields. This in turn implies that the fibers are flat submanifolds of $\mathbb{R}^{n+k}$. From the above description of the action ψ, they can be viewed as generalized helices.

The soul argument no longer works of course for metric foliations, since one has no global complete quotient space. Using different methods, it was shown in [62] that they are also homogeneous, at least for leaves of dimension less than three.

In [24], Boltner studies the so-called *equidistant* foliations of Euclidean space. These are singular metric foliations in the sense that leaves need not share the same dimension, but on the other hand, they are required to be imbedded submanifolds, and furthermore globally equidistant; i.e., the distance function from a fixed leaf is constant when restricted to a leaf. The latter condition guarantees that the space B of leaves inherits a metric space structure, and is in fact an Alexandrov space of nonnegative curvature as defined for example in [29], with the projection $\mathbb{R}^{n+k} \to B$ a submetry. Just as in the fibration case, B is shown to have a one-point set that is totally convex, the preimage of which is an affine subspace. The foliation is not, however, necessarily homogeneous.

4.3 Metric foliations of spheres

We now consider a k-dimensional metric foliation $\mathcal{F}$ of the Euclidean sphere $M = S^{n+k}$. All local results and most global ones actually hold on any complete space form of positive curvature, since such a folation can be lifted to the universal cover. Nevertheless, we shall assume for the sake of simplicity that M is a sphere.

According to Theorem 1.8.1, there is a single dual leaf, so that the dual distribution at any point consists of the whole tangent space. This suggests that the A-tensor is highly nontrivial. We begin with the following

Lemma 4.3.1. *If x is a nonzero horizontal vector, then $A_x^* u \neq 0$ for any eigenvector u of S_x.*

Proof. If not, then the holonomy field J along $t \mapsto \gamma_x(t) := \exp(tx)$ that equals u when $t = 0$ satisfies $J'(0) = -S_x J(0) - A_x^* J(0) = -\lambda J(0)$ for some scalar λ. Then $J = (\cos - \lambda \sin)E$, where E is the parallel field along γ_x with $E(0) = u$. This contradicts the fact that J can never vanish. □

As a consequence, the A-tensor cannot vanish at any single point of M.

Definition 4.3.1. $\mathcal{F}$ is said to be *substantial* along a leaf L if there exists a normal vector $x \in \mathcal{H}_p$ at some $p \in L$ such that $A_x : \mathcal{H}_p \to \mathcal{V}_p$ is onto, or equivalently, if A_x^* is one-to-one.

Of course, if A_x^* is one-to-one, then it remains so for all x in an open dense subset of $\mathcal{H}_p$. Furthermore, this condition is independent of the point p in L, since $A_X Y$ has constant norm along L for basic X, Y by O'Neill's curvature formula. Now, Theorem 1.5.1 implies in our present context that

$$(\nabla_z^{\mathrm{v}} A)_x y = S_z A_x y - S_y A_z x - S_x A_y z, \qquad x, y, z \in \mathcal{H}. \tag{4.3.1}$$

In particular, if $x = z = \dot{c}(t)$ is the tangent field of a horizontal geodesic c, and Y is horizontally parallel along c, then

$$(A_{\dot{c}} Y)'^{\mathrm{v}} = 2 S_{\dot{c}} A_{\dot{c}} Y,$$

so that the kernel of $A_{\dot{c}}$ is horizontally parallel, and $A_{\dot{c}}$ has constant rank. Thus, if $\mathcal{F}$ is substantial along a leaf L, then it remains so along all leaves in an open, dense subset of M.

Proposition 4.3.1. *If $k \leq 3$, then $\mathcal{F}$ is substantial everywhere.*

Proof. Although the argument requires considering several cases (and is therefore fairly long), it always relies in an essential way on Lemma 4.3.1. Let $p \in M$, L the leaf through p. We may assume that $S_x \neq 0$ for any nonzero x, for otherwise the claim follows from Lemma 4.3.1. Thus, the linear map $x \mapsto S_x$ from $\mathcal{H}_p$ to the space of self-adjoint endomorphisms of $\mathcal{V}_p$ is one-to-one, and in particular, $n \leq k(k+1)/2$. On the other hand, $n+k$ must be odd – the tangent bundle of an even-dimensional sphere admits no proper subbundles – so the only remaining possibilities are (k,n) equaling (2,3), (3,6), (3,4), (2,1), or (3,2). In the first three cases, where $n \geq k$, consider, for $u \in \mathcal{V}_p = \mathbb{R}^k$, the skew-adjoint endomorphism A_u of $\mathcal{H}_p = \mathbb{R}^n$ given by $A_u x = A_x^* u$ for $x \in \mathbb{R}^n$. We claim that for any nonzero u,

$$\operatorname{rank} A_u > n - k. \tag{4.3.2}$$

In particular, A_u is nonzero if $u \neq 0$, so that

$$\dim E = k, \qquad E = \{A_u \mid u \in \mathbb{R}^k\}. \tag{4.3.3}$$

Thus, $\mathcal{V}_p = \mathbb{R}^k$ is spanned by all $A_x y$, $x, y \in \mathcal{H}_p$. To see why (4.3.2) holds, assume to the contrary that A_v has rank $\leq n-k$ for some nonzero $v \in \mathcal{V}_p$; then A_v has nullity $\geq k$, and the space $W_v = \{S_x \mid x \in \ker A_v\}$ has dimension at least k by Lemma 4.3.1 again. But W_v must then intersect the space of self-adjoint endomorphisms of $\mathcal{V}_p$ that have v as eigenvector, since the latter, as a subspace of the space of all self-adjoint endomorphisms, has codimension $k-1$. In other words, there exists a nonzero x such that v is an eigenvector of S_x and $A_x^* v = 0$, contradicting Lemma 4.3.1.

An equivalent way of saying that $\mathcal{F}$ is substantial along L is that there exists a vector $x \in \mathcal{H}_p$ that is not annihilated by any nonzero element of E from (4.3.3); i.e., $A_u x \neq 0$ for any nonzero $u \in \mathcal{V}_p$. The case $(k,n) = (2,3)$ then follows, since a two-dimensional space E of skew-adjoint endomorphisms of $\mathbb{R}^3$ cannot annihilate all of $\mathbb{R}^3$. Although this can easily be argued directly, we will prove it instead in the setting that will be used in the other cases: consider the real projective space $\mathbb{R}P^2$ of the three-dimensional vector space $\mathfrak{o}(3)$ of all skew-adjoint endomorphisms of $\mathbb{R}^3$; since any nonzero element of $\mathfrak{o}(3)$ has nullity 1, the subset $\bar{E}$ of $\mathbb{R}P^2 \times \mathbb{R}^3$ consisting of all $([\alpha], u)$, where $\alpha \in E \setminus \{0\}$ and $u \in \ker \alpha$, is a smooth line bundle over a curve in $\mathbb{R}P^2$. The projection $\pi_2 : \bar{E} \to \mathbb{R}^3$ onto the second factor has as image the set of points in $\mathbb{R}^3$ annihilated by E, and the latter has therefore measure zero.

Next, let $k = 3$ and $n = 6$. By (4.3.2), any nonzero $\alpha \in E$ has nullity at most 2; thus, any given α is either invertible or has two-dimensional kernel. If no α is invertible, then as above, the subset $\bar{E}$ of $\mathbb{R}P^5 \times \mathbb{R}^6$ consisting of all pairs $([\alpha], u)$ with $\alpha \in E \setminus \{0\}$ and $u \in \ker \alpha$ is a plane bundle over a two-dimensional projective space, and E cannot annihilate a set of dimension greater than 4. So assume some $\alpha \in E$ is invertible. Recall the canonical isomorphism $\Lambda_2(\mathbb{R}^{2n}) \cong \mathfrak{o}(2n)$ that maps $u \wedge v$ to the skew-adjoint transformation $w \mapsto \langle v, w \rangle u - \langle u, w \rangle v$, and let $\bar{\alpha} \in \Lambda_2(\mathbb{R}^{2n})$ denote the bivector associated to $\alpha \in \mathfrak{o}(2n)$. Notice that α is singular iff $f(\alpha) = 0$, where f is the Pfaffian, $f(\alpha) = \star\bar{\alpha}^n/n$. Thus, $f(\alpha)$ is a homogeneous cubic polynomial in the components of α relative to any given basis of E, and the annihilating set $f^{-1}(0)$ is a cone over a manifold of dimension ≤ 1. The set $\bar{E}$ above is then a plane bundle over this manifold, and cannot annihilate all of $\mathbb{R}^6$.

We next consider the case $k = 3$ and $n = 4$. If the Pfaffian is not identically zero, then the claim follows as above, so we only need to show that f cannot be trivial. In that situation,

$$0 = 2f(\alpha) = \star\bar{\alpha} \wedge \bar{\alpha}, \qquad \alpha \in E,$$

(i.e., $\bar{\alpha}$ is decomposable), and by polarization, $\bar{\alpha} \wedge \bar{\beta} = 0$ for any α, $\beta \in E$. Consider a basis α_i of E, $1 \leq i \leq 3$. Since $\bar{\alpha}_1 \wedge \bar{\alpha}_2 = 0$, they share a common factor, and we may write

$$\bar{\alpha}_1 = \epsilon_0 \wedge \epsilon_1, \qquad \bar{\alpha}_2 = \epsilon_0 \wedge \epsilon_2$$

for some independent one-forms ϵ_0, ϵ_1, ϵ_2 on $\mathbb{R}^4$. Now, $\bar{\alpha}_3$ may or may not lie in the span of $\epsilon_i \wedge \epsilon_j$, $0 \leq i < j \leq 2$. In the former case, consider any ϵ_3 that does not belong to the span of ϵ_i, $0 \leq i \leq 2$. If e_i denotes the basis dual to ϵ_i, then all of E annihilates e_3, which contradicts Lemma 4.3.1. In the latter case, $\bar{\alpha}_3 = \beta \wedge \epsilon_3$, and since it shares a common factor with α_1 and α_2,

$$\bar{\alpha}_3 = (s_0\epsilon_0 + s_1\epsilon_1) \wedge \epsilon_3 = (t_0\epsilon_0 + t_2\epsilon_2) \wedge \epsilon_3.$$

It follows that $s_0 = t_0$, and $s_1 = t_2 = 0$; i.e., $\bar{\alpha}_3$ is a multiple of $\epsilon_0 \wedge \epsilon_3$, and no nonzero element of E annihilates the vector e_0 of the dual basis.

Finally, the last two cases cannot occur by [90] (cf. also [91]), where Molino provides a classification of Riemannian foliations of codimension $k \leq 3$ on compact, simply connected manifolds. In our situation, this also follows by a direct argument: the case $(k, n) = (2, 1)$ may be ruled out since otherwise $A \equiv 0$, contradicting Lemma 4.3.1. Next, consider $k = 3$, $n = 2$. At any point, the image of the A-tensor is one-dimensional, and the claim again follows from Lemma 4.3.1, if we can establish that for some nonzero x, S_x has an eigenvector orthogonal to that image; i.e., if given any two-dimensional subspace E of self-adjoint endomorphisms of $\mathbb{R}^3$ and any plane P through 0 in $\mathbb{R}^3$, some element in $E^* = E \setminus \{0\}$ has an eigenvector in P. We will argue this by contradiction: if not, then each element of E^* has three distinct eigenvalues, thus defining continuous functions $\lambda_i : E^* \to \mathbb{R}$ with $\lambda_1 < \lambda_2 < \lambda_3$. Similarly, we can find continuous unit eigenvector fields $X_i : E^* \to \mathbb{R}^3 \setminus \{0\}$, $SX_i(S) = \lambda_i(S)X_i(S)$ for $S \in E^*$, with image contained in one of the two open half-spaces with boundary P. But $-S$ has eigenvalues $-\lambda_1(S) > -\lambda_2(S) > -\lambda_3(S)$, so that $X_1(-S) = X_3(S)$, $X_2(-S) = X_2(S)$, and $X_3(-S) = X_1(S)$. Thus, $X_1 \wedge X_2 \wedge X_3(-S) = -X_1 \wedge X_2 \wedge X_3(S)$, which is impossible since E^* is connected. □

From now on, we assume, unless otherwise specified, that the leaf dimension k is no larger than 3. Let U denote a connected open set such that the restriction

$\mathcal{F}_{|U}$ is given by the fibers of a Riemannian submersion $\pi : U \to B$, and consider the space $\mathfrak{A}$ of *integrability fields* spanned by all $A_X Y$ on U where X, Y are elements of the space $\mathfrak{B}$ of basic fields on U. Our next endeavor is to show that $\mathfrak{A}$ is a Lie algebra. Notice first that by Proposition 1.5.1,

$$\pi_* A_X^* A_X Y = \frac{1}{3}\big(\pi_* R(X,Y)X - R^B(\pi_* X, \pi_* Y)\pi_* X\big),$$

so that $A_X^* A_X Y \in \mathfrak{B}$ if X, $Y \in \mathfrak{B}$, and thus,

$$T\langle A_X Y, A_X Z\rangle = 0, \qquad T \text{ vertical}, \quad X, Y, Z \in \mathfrak{B}. \tag{4.3.4}$$

Lemma 4.3.2. *If X, $Y \in \mathfrak{B}$, then $S_X A_X Y \in \mathfrak{A}$.*

Proof. (4.3.1) implies

$$2\langle S_X A_X Y, A_X Y\rangle = \frac{1}{2} X|A_X Y|^2 - \langle A_X \nabla_X^{\mathbf{h}} Y, A_X Y\rangle - \langle A_Y \nabla_X X, A_Y X\rangle,$$

which is constant along leaves by (4.3.4), since $TX = XT - [T, X]$, and $[T, X]$ is vertical. By polarization,

$$T\langle S_X A_X Y, A_X Z\rangle = 0, \qquad T \text{ vertical}, \quad X, Y, Z \in \mathfrak{B}. \tag{4.3.5}$$

Consider a leaf L in U. Since $\mathcal{F}$ is substantial, we may assume that A_X is onto L. Using (4.3.4) and (4.3.5), we can find $Y_1, \dots, Y_k \in \mathfrak{B}$ such that $A_X Y_{i|L}$ is an orthonormal frame of eigenvector fields of S_X with constant eigenvalues λ_i along L. Then for any basic Y, the restriction $A_X Y_{|L}$ is a constant linear combination $\sum_i \alpha_i A_X Y_i$, with $\alpha_i = \langle A_X Y, A_X Y_i\rangle$, and $S_X A_X Y = A_X Z$, where $Z = \sum_i \alpha_i \lambda_i Y_i \in \mathfrak{B}$. Thus, $S_X A_X Y \in \mathfrak{A}$. □

Proposition 4.3.2. *$\mathfrak{A} \oplus \mathfrak{B}$ is a Lie algebra that contains $\mathfrak{A}$ as an ideal.*

Proof. For X, $Y \in \mathfrak{B}$, $T \in \mathfrak{A}$, the Jacobi identity implies

$$\begin{aligned} 2[A_X Y, T] &= [[X,Y]^{\mathbf{v}}, T] = [[X,Y], T] - [[X,Y]^{\mathbf{h}}, T] \\ &= [X, [Y,T]] - [Y, [X,T]] - [[X,Y]^{\mathbf{h}}, T], \end{aligned}$$

and it remains to show that

$$[Y, T] \in \mathfrak{A}, \qquad Y \in \mathfrak{B}, T \in \mathfrak{A}. \tag{4.3.6}$$

Now, by (4.3.1),

$$[X, A_X Y] = \nabla_X^{\mathbf{v}} A_X Y + S_X A_X Y = 3 S_X A_X Y + A_X \nabla_X^{\mathbf{h}} Y - A_Y \nabla_X^{\mathbf{h}} X,$$

and using Lemma 4.3.2, we conclude that $\big[X, [X,Y]^{\mathbf{v}}\big] \in \mathfrak{A}$. Thus, by polarization,

$$\big[X, [Y,Z]^{\mathbf{v}}\big] + \big[Y, [X,Z]^{\mathbf{v}}\big] \in \mathfrak{A}. \tag{4.3.7}$$

Furthermore,

$$[Y,[Z,X]^{\mathbf{v}}]+[Y,[X,Z]^{\mathbf{v}}]=0, \tag{4.3.8}$$

and

$$[Z,[X,Y]^{\mathbf{v}}]+[Y,[X,Z]^{\mathbf{v}}]=-[Z,[Y,X]^{\mathbf{v}}]-[Y,[Z,X]^{\mathbf{v}}]\in\mathfrak{A} \tag{4.3.9}$$

by (4.3.7). Adding (4.3.7) through (4.3.9) then implies

$$(\circlearrowleft[X,[Y,Z]^{\mathbf{v}}])+3[Y,[X,Z]^{\mathbf{v}}]\in\mathfrak{A}, \tag{4.3.10}$$

where $\circlearrowleft$ denotes cyclic summation. Now, $[X,[Y,Z]^{\mathbf{v}}]$ is vertical by the Jacobi identity, so that

$$\begin{aligned}\circlearrowleft[X,[Y,Z]^{\mathbf{v}}] &= \circlearrowleft[X,[Y,Z]]-\circlearrowleft[X,[Y,Z]^{\mathbf{h}}]=-\circlearrowleft[X,[Y,Z]^{\mathbf{h}}]^{v}\\ &=-2\circlearrowleft A_X[Y,Z]^{\mathbf{h}}\in\mathfrak{A},\end{aligned}$$

which, together with (4.3.10), proves (4.3.6). □

It follows from Proposition 4.3.2 that the restriction $\mathfrak{A}_L$ of $\mathfrak{A}$ to a leaf L in U is a Lie algebra with dimension $k\le\dim\mathfrak{A}_L\le\binom{n}{2}$. We now improve on this estimate:

Lemma 4.3.3. *$\langle T_1,T_2\rangle$ is constant along L for any $T_i\in\mathfrak{A}_L$. In particular, $\mathfrak{A}_L$ has dimension k.*

Proof. It must be shown that $\langle A_{Z_1}Z_2,A_{Z_3}Z_4\rangle$ is constant along L for any $Z_i\in\mathfrak{B}$. Since A_X is onto $\mathcal{V}_L$ for an open dense subset of basic fields X along L, we may assume that the Z_i belong to a subspace H of basic fields along L, of dimension $3\le\dim H=m+1\le 4$, such that $A_{X_0}(H)=\mathcal{V}_L$ for some $X_0\in H$. By (4.3.4), there exist linearly independent $X_1,\dots,X_m$ such that $\{A_{X_0}X_i\mid i\le k\}$ is an orthonormal basis of $\mathcal{V}_L$. Using skew-symmetry of A, it suffices to show that $\langle A_{X_i}X_j,A_{X_0}X_l\rangle$ is constant for $0\le i<j\le m$, $1\le l\le k$. Now, by (4.3.4), this is true if $i=0$, or $i=l$, or $j=l$. The other cases then follow from (4.3.4) together with the fact that $A_{X_i}X_j$ has constant length: for example, when $k=3$, then

$$\langle A_{X_1}X_2,A_{X_0}X_3\rangle^2=|A_{X_1}X_2|^2-\langle A_{X_1}X_2,A_{X_0}X_1\rangle^2-\langle A_{X_1}X_2,A_{X_0}X_2\rangle^2$$

is constant. □

We are now in a position to prove the main result of this section. The argument will make use of the following classical theorem, a proof of which can be found in [116]:

Theorem 4.3.1 (Fundamental theorem for submanifolds). *Let M_i, $i=1,2$, denote k-dimensional Riemannian submanifolds of the simply connected spaceform Q_c^{n+k} of constant curvature c, $h:M_1\to M_2$ an isometry. Let $E(\nu_i)$ denote the total space of the normal bundle ν_i of M_i in Q_c, and suppose there exists a linear* *bundle isometry $H : E(\nu_1) \to E(\nu_2)$ covering h, such that H preserves the normal connections $\nabla_i^{\mathbf{h}}$ on ν_i and the second fundamental forms S^i of M_i; i.e.,*

$$\nabla_2^{\mathbf{h}}{}_T(HX) = H\nabla_1^{\mathbf{h}}{}_T X, \qquad h_* S_X^1 T = S_{HX}^2 h_* T$$

for any sections T and X of the tangent and normal bundle respectively of M_1. Then there exists an isometry $\tilde{h}$ of Q_c such that $\tilde{h}_{|M_1} = h$, and the restriction of $\tilde{h}_*$ to $E(\nu_1)$ equals H.

Theorem 4.3.2 (Gromoll-Grove, [56]). *When $k \leq 3$, any k-dimensional metric foliation of the Euclidean sphere S^{n+k} is homogeneous; specifically, it is the orbit foliation of a connected k-dimensional Lie subgroup of $SO(n+k+1)$.*

Proof. Consider a point p in the sphere, and the leaf L containing it. We begin by constructing a group of local isometries of L near p. These will then be extended to the whole ambient space. Denote by G the local Lie group of diffeomorphisms of some neighborhood of p in L generated by the flows of vector fields in $\mathfrak{A}_L$. There are neighborhoods U of e in G and V of p in L such that $\imath_p : U \to V$, $\imath_p(g) := g(p)$, is a diffeomorphism. According to the discussion in Section 2.3, a vector field on V belongs to $\mathfrak{A}_L$ iff it is $\imath_p$-related to a right invariant vector field of G. Denote by $\mathfrak{K}_L$ the algebra of vector fields on V that are $\imath_p$-related to *left* invariant vector fields of G; i.e.,

$$\mathfrak{K}_L = \{T \in \mathfrak{X}(M) \mid T = \imath_{p*} X \circ \imath_p^{-1},\ X \in \mathfrak{g}\}.$$

Since left and right invariant fields commute, $[\mathfrak{A}_L, \mathfrak{K}_L] = 0$. This implies that $\mathfrak{K}_L$ is an algebra of Killing fields: in fact, since $\mathfrak{A}_L$ contains a point-wise orthonormal basis of the vertical space, it suffices to check that the transformation $T \mapsto \nabla_T X$, $X \in \mathfrak{K}_L$, is skew-adjoint on these basis elements. But this is clear, since

$$\langle \nabla_T X, T\rangle = \langle \nabla_X T, T\rangle = \frac{1}{2} X\langle T, T\rangle = 0, \quad T \in \mathfrak{A}_L.$$

We next extend the isometries of $V \subset L$ generated by $\mathfrak{K}_L$ to (unique) leaf-preserving isometries of an open set in the sphere. Using the fact that a local Killing field has a unique global extension, the theorem then clearly follows. So consider such a local isometry ϕ, and extend it to a linear isometry Φ of the normal bundle of L near p by defining

$$\Phi X := X \circ \phi, \quad X \in \mathfrak{B}.$$

We claim that Φ preserves the normal connection: if $T \in \mathfrak{A}$ and $X \in \mathfrak{B}$, then $\nabla_T^{\mathbf{h}} X = -A_X^* T$ is basic by Lemma 4.3.3, and $\phi_* T = T \circ \phi$ because the algebras $\mathfrak{A}$ and $\mathfrak{K}_L$ commute. Thus,

$$\Phi(\nabla_T^{\mathbf{h}} X) = (\nabla_T^{\mathbf{h}} X) \circ \phi = \nabla_{T\circ\phi}^{\mathbf{h}} X = \nabla_{\phi_* T}^{\mathbf{h}} X = \nabla_T^{\mathbf{h}}(X \circ \phi) = \nabla_T^{\mathbf{h}}(\Phi X).$$

In the same way, Φ preserves the second fundamental form: Lemmas 4.3.2 and 4.3.3 imply that $S_X \mathfrak{A} \subset \mathfrak{A}$, so that

$$\phi_* S_X T = (S_X T) \circ \phi = S_{X\circ\phi}(T \circ \phi) = S_{\Phi X} \phi_* T.$$

The fundamental theorem for submanifolds then implies that ϕ extends to an isometry of a tubular neighborhood of V in the ambient space. Since this isometry must then locally be given by $\exp_V \circ \Phi \circ \exp_V^{-1}$, where $\exp_V$ is the exponential map of the normal bundle of V, it preserves leaves. □

Little is known at present concerning metric foliations of spheres with higher-dimensional leaves. One remarkable fact is that when $k > 1$, they are always generalized Seifert fibrations, in the sense that all leaves are compact [52]. The latter are fairly similar to actual fibrations, at least from a homotopical point of view [75].

We end the section with a brief description of the foliations that can occur in Theorem 4.3.2. If $k = 1$, then $\mathcal{F}$ is the orbit foliation of a one-dimensional Lie subgroup G of $SO(n+2)$, and is therefore determined by a homomorphism $\phi : \mathbb{R} \to SO(n+2)$, $\phi(t) = e^{tM}$, where $M = \dot\phi(0) \in \mathfrak{o}(n+2)$. The skew-symmetric matrix M is similar, via an orthogonal matrix, to a block matrix of the form

$$\operatorname{diag}\{i\lambda_1, \dots, i\lambda_s, 0 \dots, 0\} := \begin{bmatrix} 0 & -\lambda_1 & & & & & & \\ \lambda_1 & 0 & & & & & & \\ & & \ddots & & & & & \\ & & & 0 & -\lambda_s & & & \\ & & & \lambda_s & 0 & & & \\ & & & & & 0 & & \\ & & & & & & \ddots & \\ & & & & & & & 0 \end{bmatrix}$$

where $0 < \lambda_1 \le \cdots \le \lambda_s$. Since the action is free, M must actually have the form $\operatorname{diag}\{i\lambda_1, \dots, i\lambda_s\}$, and n is even, with $s = 1 + (n/2)$. Normalize M so that $\lambda_s = 1$. Then, up to congruence, G is a direct sum of rotations

$$\operatorname{diag}\{e^{i\lambda_1 t}, e^{i\lambda_2 t}, \dots, e^{it}\}, \qquad 0 < \lambda_1 \le \lambda_2 \le \cdots \le 1.$$

Notice there are always at least s compact leaves that are totally geodesic circles, namely the orbits of the odd standard basis vectors $\mathbf{e}_1, \mathbf{e}_3, \dots, \mathbf{e}_{n+1}$. All leaves are compact iff each λ_j is rational. Among these foliations, only one is a fibration, namely the Hopf fibration, corresponding to $\lambda_j = 1$ for all j.

Next, consider the case $k = 2$. Since the only two-dimensional subgroups of an orthogonal group are abelian, there can be no metric foliations of this dimension: for otherwise, there would exist independent M, N in the Lie algebra of G with vanishing bracket. Then M and N would share a common basis of complex eigenvectors, and such a vector would have the same orbit under the actions $t \mapsto e^{tM}$, $t \mapsto e^{tN}$, implying the action is not free. This also shows, incidentally, that there are no free $O(k)$-actions on spheres if $k > 3$, since the orthogonal Lie algebra is then no longer simple, and contains linearly independent commuting vectors.

When $k = 3$, the last argument implies that G has $SU(2)$ as universal cover. The classification in this case is obtained via representation theory, and we will limit ourselves to merely stating the result. The interested reader should consult [56] and [27] for further details. Let V_n denote the complex vector space of homogeneous polynomials p of degree n in two complex variables z_1, z_2,

$$p(z_1, z_2) = \sum_{k=0}^{n} a_k z_1^k z_2^{n-k},$$

and define an action ρ_n of $SU(2)$ on V_n by setting

$$(gp)(z) = p(zg), \qquad g \in SU(2), \quad p \in V_n, \quad z = (z_1, z_2),$$

with zg denoting matrix multiplication. Notice that ρ_1 is just the standard action of $SU(2)$ on $V_1 \cong \mathbb{C}^2$. The main result is that three-dimensional foliations of S^{n+3} exist precisely when $n = 4l$, and any such foliation is given by a direct sum $\rho_{n_1} \oplus \cdots \oplus \rho_{n_s}$ of irreducible representations of $SU(2)$, with n_j odd for all j, $1 \leq n_1 \leq \cdots \leq n_s$, and $n_1 + \cdots + n_s = 2((n/4) + 1) - s$. Here again, only one is a fibration, namely the Hopf fibration with $n_1 = \cdots = n_s = 1$.

As far as metric *fibrations* of spheres are concerned, it follows from [28] that the fiber must be a homotopy sphere of dimension one, three, or seven. The first two cases are covered in Theorem 4.3.2, and the last one was solved by Wilking in [139], using Morse theoretical methods:

Theorem 4.3.3 (Gromoll-Grove, Wilking). *Any Riemannian submersion $S^{n+k} \to M^n$ of a Euclidean sphere is congruent to a Hopf fibration.*

In the special case of totally geodesic fibers, this result is due to Escobales [46] and Ranjan [107]. The extra assumption is quite strong, of course, and it is easy to see directly that in this case, M must be a rank one symmetric space: consider a point p in M. Local geodesic reflection in p of a curve c can be obtained by horizontally lifting that curve to the sphere, reflecting it in the fiber over p, and projecting back onto M. The first two steps preserve the length of c since the fiber is totally geodesic, so that geodesic reflection in M is distance non-increasing. It must then be an isometry, because its square is the identity. Thus, M is locally symmetric. If the fiber of π is connected, then M is simply connected, and hence globally symmetric. The rank statement follows from the fact that M has positive curvature by O'Neill's formula.

The discussion of foliations in space forms carried out in the last two sections raises several new questions: it would for example be interesting to determine how much of the rigidity that is apparent in constant nonnegative curvature carries over to more general manifolds, such as symmetric spaces. Most of the few known results deal with one-dimensional metric foliations: they have been shown to be homogeneous if the ambient space is a compact Lie group [94] with bi-invariant metric or $S^2 \times \mathbb{R}$ with the standard product metric [61]. The methods used in each case are specific to the situation at hand and do not easily generalize. In a related but slightly different direction, it is known that the same result holds for the Heisenberg group [95]; there are, however, noncompact Lie groups with left invariant metrics that admit one-dimensional metric foliations which are not homogeneous [135].

4.4 Geometry of the tangent bundle

In order to discuss metric foliations on a compact space form M of nonpositive curvature, some properties of the geodesic flow on the tangent bundle of M will be needed. The reader familiar with these concepts may skip this section without loss of continuity, and the one who wishes to study them in more detail is referred to [102] or [106].

Denote by $\pi : TM \longrightarrow M$ and $\tilde{\pi} : T^*M \longrightarrow M$ the bundle projections.

Definition 4.4.1. The fundamental 1-form θ on the co-tangent bundle T^*M is given by $\theta(\alpha) = \tilde{\pi}^*\alpha$, for $\alpha \in T^*M$.

Thus, for $\xi \in (T^*M)_\alpha$, $\theta(\alpha)(\xi) = \alpha(\tilde{\pi}_{*\alpha}\xi)$.

The Levi-Civita connection $\mathcal{H}$ of a Riemannian manifold M induces a bundle homomorphism $K : TTM \longrightarrow TM$ over $\pi : TM \longrightarrow M$, called the *connection map*, defined as follows: a vector $\xi \in TTM$ decomposes as $\xi^{\mathbf{h}} + \xi^{\mathbf{v}} \in \mathcal{H} \oplus \mathcal{V}$, where $\mathcal{V} = \ker \pi_*$ and $\mathcal{H} = \ker K$. For $p \in M$, $u \in M_p$, and $\mathcal{V}_u = \imath_{*u}(M_p)$, with $\imath : M_p \hookrightarrow TM$ denoting inclusion, denote by $\mathcal{J}_u : M_p \longrightarrow (M_p)_u$ the isomorphism given by $\mathcal{J}_u w = \dot{\gamma}(0)$, $\gamma(t) = u + tw$. Then

$$K\xi = (\imath_* \circ \mathcal{J}_u)^{-1}\xi^{\mathbf{v}}. \tag{4.4.1}$$

Alternatively, for a vector field X on M and $u \in TM$,

$$\nabla_u X = K X_* u. \tag{4.4.2}$$

Since the restrictions $\pi_* : \mathcal{H}_u \longrightarrow M_{\pi(u)}$ and $K : \mathcal{V}_u \longrightarrow M_{\pi(u)}$ are both isomorphisms, the map

$$(\pi_*, K) : TTM \longrightarrow TM \oplus TM$$

is a bundle isomorphism over $\pi : TM \longrightarrow M$. In fact, its inverse $\mathcal{I} : TM \oplus TM \longrightarrow TTM$ is described as follows: for $u \in TM$, $w, z \in M_{\pi(u)}$, consider a curve $\gamma : I \to M$ with $\dot{\gamma}(0) = z$. If Z denotes the parallel field along γ with $Z(0) = u$ (i.e., Z is the horizontal lift to TM of γ starting at u), then

$$\mathcal{I}(z, w) = \dot{Z}(0) + \imath_* \mathcal{J}_u w.$$

We will routinely identify $(TM)_u$ with $M_{\pi(u)} \times M_{\pi(u)}$ via the isomorphism (π_*, K). The *Sasaki metric* $\langle\langle , \rangle\rangle$ on the manifold TM is that metric for which (π_*, K) becomes a linear isometry. It is a connection metric in the sense of Proposition 2.7.1.

Recall that a vector field S on TM is called a *spray* on M if $\pi_* \circ S = 1_{TM}$ and $S \circ \mu_a = a\mu_{a*}S$, where μ_a denotes multiplication by $a \in \mathbb{R}$. The *geodesic spray* S is the unique horizontal spray on M; i.e., $S(u) = (u, 0)$, $u \in TM$. The integral curves of S are precisely the velocity fields $\dot{\gamma} : I \longrightarrow TM$ of geodesics $\gamma : I \longrightarrow M$ of M.

We shall denote by $\flat : TM \longrightarrow T^*M$, $\flat(u) = \langle u, \cdot \rangle$, and $\tilde{\flat} : TTM \longrightarrow T^*TM$ the musical isomorphisms induced by the metrics on M and TM respectively. The cotangent vector $\flat(u)$ is often written as $u^\flat$. The next proposition says that the geodesic spray is essentially the metric dual of the fundamental 1-form on T^*M:

Proposition 4.4.1. $S^{\tilde{\flat}} = \flat^*\theta$.

Proof. Let $u \in TM$, $\xi \in (TM)_u$. Since $\tilde{\pi} \circ \flat = \pi$ and S is horizontal,

$$\begin{aligned} \flat^*\theta(\xi) &= \theta(\flat_*\xi) = u^\flat(\tilde{\pi}_* \circ \flat_*\xi) = u^\flat(\pi_*\xi) = \langle u, \pi_*\xi \rangle = \langle \pi_* S(u), \pi_*\xi \rangle \\ &= \langle\langle S(u), \xi \rangle\rangle. \end{aligned}$$ □

Define a complex structure J on TTM by setting $J(u, w) = (-w, u)$; equivalently,

$$\pi_* J = -K, \qquad KJ = \pi_*. \tag{4.4.3}$$

Then the 2-form Ω, where

$$\Omega(\xi, \eta) := \langle\langle J\xi, \eta \rangle\rangle,$$

is a symplectic (i.e., nondegenerate) 2-form on TM, and if n is the dimension of M, then Ω^n equals $(-1)^{[n/2]}n!$ times the Sasaki metric volume element. On the other hand, $-d\theta$ is also a symplectic form, but one on T^*M rather than TM. The relation between the two is given by the following:

Proposition 4.4.2. $\Omega = -d(b^*\theta)$; *i.e.,* Ω *is the metric pullback of the canonical symplectic form* $-d\theta$ *on* T^*M.

Proof. Viewing the identity map 1_{TM} on TM as a vector field on M along $\pi : TM \longrightarrow M$, we have

$$\nabla_X(1_{TM}) = K(1_{TM})_*X = KX, \qquad X \in \mathfrak{X}TM.$$

Thus, if Y is another vector field on TM, then

$$\begin{aligned} -d(\flat^*\theta)(X, Y) &= -X\langle \pi_* S, \pi_* Y \rangle + Y\langle \pi_* S, \pi_* X \rangle + \langle \pi_* S, \pi_*[X, Y] \rangle \\ &= -\langle \nabla_X(1_{TM}), \pi_* Y \rangle - \langle 1_{TM}, \nabla_X \pi_* Y \rangle + \langle \nabla_Y(1_{TM}), \pi_* X \rangle \\ &\quad + \langle 1_{TM}, \nabla_Y \pi_* X \rangle + \langle 1_{TM}, \pi_*[X, Y] \rangle \\ &= -\langle KX, \pi_* Y \rangle + \langle KY, \pi_* X \rangle = \langle \pi_* JX, \pi_* Y \rangle + \langle KJX, KY \rangle \\ &= \langle\langle JX, Y \rangle\rangle, \end{aligned}$$

where we used (4.4.3) in the equality before last. □

Proposition 4.4.3. *If $h : TM \longrightarrow \mathbb{R}$ denotes the* energy function, $h(u) = (1/2)|u|^2$, *then $i_S\Omega = dh$.*

Proof. S is horizontal for the submersion $\pi : TM \longrightarrow M$, and if γ is an integral curve of S, then $\pi \circ \gamma$ is a geodesic of M. Thus, γ is a geodesic of the Sasaki metric, and S is an auto-parallel vector field. Given a vector field X on TM,

$$\begin{aligned} i_S\Omega(X) &= -d(\flat^*\theta)(S,X) = -S\langle\langle S,X\rangle\rangle + X\langle\langle S,S\rangle\rangle + \langle\langle S,[S,X]\rangle\rangle \\ &= -\langle\langle S,\nabla_S X - [S,X]\rangle\rangle + X\langle\langle S,S\rangle\rangle = -\langle\langle S,\nabla_X S\rangle\rangle + X\langle\langle S,S\rangle\rangle \\ &= \frac{1}{2}X\langle\langle S,S\rangle\rangle = X(h), \end{aligned}$$

since $\langle\langle S,S\rangle\rangle(u) = \langle\pi_*S,\pi_*S\rangle(u) = \langle u,u\rangle$. □

Proposition 4.4.3 says that the geodesic spray is the *Hamiltonian* vector field of the energy function with respect to Ω.

Assume from now on that M is compact. Instead of working on TM, we shall restrict ourselves to the unit tangent bundle $T^1M = \{u \in TM \mid |u| = 1\}$, which has the advantage of being compact. We first describe the tangent space of this manifold at a given point:

Proposition 4.4.4. *If $\imath : T^1M \hookrightarrow TM$ denotes the inclusion map, then for $u \in T^1M$,*

$$\imath_*(T^1M)_u = \{\xi \in (TM)_u \mid \langle K\xi, u\rangle = 0\} = J \circ S(u)^\perp.$$

Alternatively, under the isomorphism (π_*, K), $\imath_*(T^1M)_u = (0,u)^\perp$. In particular, there is a unique vector field on T^1M that is $\imath$-related to S (it will be denoted by S also).

Proof. Since T^1M is the pre-image of 1 under the energy function h, the space $\imath_*(T^1M)_u$ is the kernel of h_*u, which by Proposition 4.4.3 equals $\{\xi \in (TM)_u \mid \Omega(S(u),\xi) = 0\}$. But $\Omega(S(u),\xi) = \langle\langle JS(u),\xi\rangle\rangle = \langle K\xi,u\rangle$ by (4.4.3). □

We will denote by σ the restriction $\imath^*\flat^*\theta$ to T^1M of the 1-form $\flat^*\theta$ on TM. By Proposition 4.4.1, σ is the metric dual of the geodesic spray S on T^1M. Since the volume form of TM is

$$\bar{\omega} = \frac{(-1)^{[n/2]}}{n!}\Omega^n = \frac{(-1)^{n+[n/2]}}{n!}d(\flat^*\theta)^n,$$

the volume form of T^1M is

$$\omega = i_{JS}\imath^*\bar{\omega} = \frac{(-1)^{n+[n/2]}}{n!} i_{JS}d\sigma^n,$$

with i_{JS} denoting interior multiplication by JS. But $i_{JS}d\sigma^n = n(i_{JS}d\sigma)\wedge(d\sigma)^{n-1}$, and for $X \in \mathfrak{X}T^1M$,

$$i_{JS}d\sigma(X) = -\Omega(JS,X) = -\langle\langle J^2S,X\rangle\rangle = \langle\langle S,X\rangle\rangle = \sigma(X).$$

Thus,

$$\omega = \frac{(-1)^{n+[n/2]}}{(n-1)!}\sigma \wedge (d\sigma)^{n-1}. \tag{4.4.4}$$

A 1-form α on an odd-dimensional manifold M^{2n-1} is said to be a *contact form* if $\alpha \wedge (d\alpha)^{n-1}$ is nowhere zero. (4.4.4) implies that the metric dual of the geodesic spray is a contact form on the unit tangent bundle.

Since T^1M is compact, S is complete, and its flow is a one parameter group $\{\phi_t\}_{t\in\mathbb{R}}$ of diffeomorphisms, called the *geodesic flow* of M. The volume form ω has finite integral over T^1M, and thus induces a probability measure on that space, called the *Liouville measure*.

Proposition 4.4.5. *The geodesic flow is measure-preserving; i.e., given $A \subset T^1M$, the volume of $\phi_t(A)$ is constant, $t \in \mathbb{R}$.*

Proof. The statement follows once we establish that $\mathcal{L}_S\omega = 0$, or using (4.4.4), that $\mathcal{L}_S\sigma = 0$. Now, $\mathcal{L}_S\sigma = i_S d\sigma + d i_S\sigma = i_S d\sigma$, because $i_S\sigma \equiv 1$. Given $X \in \mathfrak{X}T^1M$, $i_S d\sigma(X) = -\Omega(S,X) = -\langle\langle JS, X\rangle\rangle = 0$, since JS is orthogonal to T^1M. □

Proposition 4.4.6. *Given $v \in T^1M$ and $(u,w) \in (T^1M)_v$, $\phi_{t*}(u,w) = (J(t), J'(t))$, where J is the Jacobi field along the geodesic $t \mapsto \exp(tv)$ with $J(0) = u$, $J'(0) = w$.*

Proof. Recall that for $v \in TM$, $\phi_t(v) = \dot{c}_v(t)$, where $c_v(t) = \exp(tv)$. Consider a curve $\gamma : I \longrightarrow T^1M$ with $\gamma(0) = v$, $\dot{\gamma}(0) = (u,w)$. Then

$$(t,s) \mapsto V(t,s) := \pi \circ \phi_t \circ \gamma(s) = \exp(t\gamma(s))$$

is a variation by geodesics of c_v. The corresponding Jacobi field $t \mapsto J(t) = V_*D_2(t,0)$ is given by

$$J(t) = \pi_* \circ \phi_{t*}\dot{\gamma}(0) = \pi_* \circ \phi_{t*}(u,w),$$

and

$$J'(t) = \nabla_{D_1(t,0)}V_*D_2 = \nabla_{D_2(t,0)}V_*D_1.$$

But $V_*D_1(t,s) = \phi_t(\gamma(s))$, so

$$J'(t) = \nabla_{D(0)}(\phi_t \circ \gamma) = K(\phi_t \circ \gamma)_*D(0) = K \circ \phi_{t*}(u,w),$$

as claimed. □

We end this section with two ergodic theorems that hold on measure spaces with a measure-preserving transformation, see [130]. In our context, with the transformation being the geodesic flow, they can be stated as follows:

Theorem 4.4.1. *Let A be a submanifold of the unit tangent bundle of M that is measure-invariant under the geodesic flow.*

1. (**Oseledets**) *For almost every $v \in A$, there exists a direct sum decomposition of the tangent space*

$$A_v = V^s(v) \oplus V^u(v) \oplus V^0(v)$$

of A at v, where for $\xi \neq 0$,

$$\xi \in V^s(v) \text{ iff } \lim_{t\to\pm\infty} \frac{1}{t} \ln |\phi_{t*}\xi| < 0,$$
$$\xi \in V^u(v) \text{ iff } \lim_{t\to\pm\infty} \frac{1}{t} \ln |\phi_{t*}\xi| > 0,$$
$$\xi \in V^0(v) \text{ iff } \lim_{t\to\pm\infty} \frac{1}{t} \ln |\phi_{t*}\xi| = 0.$$

2. **(Birkhoff)** *If $f : A \longrightarrow \mathbb{R}$ is integrable, then for a.e. $u \in A$,*

$$\tilde{f}(u) := \lim_{t\to\infty} \frac{1}{t} \int_0^t f(\phi_s u) ds \text{ exists, and } \int_A f\omega = \int_A \tilde{f}\omega.$$

4.5 Compact space forms of nonpositive curvature

Although at the time of writing there does not seem to be a classification of metric foliations in space forms of curvature $\kappa \leq 0$, we will see that there are severe restrictions, at least in the compact case, cf. [81], [133]. The main tools used in the argument are the ergodic theorems introduced in the last section. So let M be a compact space of constant curvature $\kappa \leq 0$, and $\mathcal{F}$ a metric foliation on M. We begin by identifying the tangent space $\mathcal{H}_x$ of the horizontal bundle $\mathcal{H}$ at $x \in \mathcal{H}$. Notice that if $\mathcal{H}^1$ denotes the unit horizontal bundle, then for $x \in \mathcal{H}^1$, $\mathcal{H}^1_x = \mathcal{H}_x \cap (0, x)^\perp$ by Proposition 4.4.4.

Lemma 4.5.1. $\mathcal{H}_x = \{(e, f) \in M_{\pi(x)} \times M_{\pi(x)} \mid f \in A_{e^{\mathbf{h}}} x - S_x e^{\mathbf{v}} + \mathcal{H}\}$.

Proof. Both spaces have the same dimension $2n - k$, so we only need to show that $\mathcal{H}_x$ is contained in the space on the right. Consider $\xi = (e, f) \in \mathcal{H}_x$ and a curve Z in $\mathcal{H}$ with $\dot{Z}(0) = \xi$. If $c := \pi \circ Z$, $p := c(0)$, then $\langle Z, U \circ c\rangle \equiv 0$ for any vertical field U, so that

$$\begin{aligned}
0 &= \langle Z, U \circ c\rangle'(0) = \langle Z', U \circ c\rangle(0) + \langle Z, (U \circ c)'\rangle(0) \\
&= \langle K\xi, U(p)\rangle + \langle x, \nabla_{\pi_\xi} U\rangle = \langle f, U(p)\rangle + \langle x, \nabla_e U\rangle \\
&= \langle f^{\mathbf{v}}, U(p)\rangle + \langle x, \nabla_{e^{\mathbf{h}}} U\rangle + \langle x, \nabla_{e^{\mathbf{v}}} U\rangle \\
&= \langle f^{\mathbf{v}}, U(p)\rangle - \langle A_{e^{\mathbf{h}}} x, U(p)\rangle + \langle S_x e^{\mathbf{v}}, U(p)\rangle.
\end{aligned}$$

Thus, $f^{\mathbf{v}} = A_{e^{\mathbf{h}}} x - S_x e^{\mathbf{v}}$, as claimed. □

Consider $\mathcal{H}^1$ as a Riemannian submanifold of T^1M, where T^1M is endowed with the Sasaki metric, and observe that $\mathcal{H}^1$ is invariant under the geodesic flow, since a geodesic that starts out horizontally remains so.

Proposition 4.5.1. *The geodesic flow $\{\phi_t\}$ is measure-preserving on $\mathcal{H}^1$.*

Proof. Given $x \in \mathcal{H}^1$ and $\xi \in \mathcal{H}^1_x$, denote by Y_ξ the Jacobi field along the geodesic $t \mapsto \exp(tx)$ with $Y_\xi(0) = \pi_*\xi$, $Y'(0) = K\xi$. By Proposition 4.4.6, $\phi_{t*}\xi = (Y_\xi(t), Y'_\xi(t))$, after the usual identification of TTM with $TM \oplus TM$ via (π_*, K).

Consider first the negative curvature case, which we normalize so that $\kappa = -1$. Then $Y_\xi(t) = e^t E_1(t) + e^{-t}E_2(t)$ for some parallel fields E_i, and given $\eta \in \mathcal{H}^1_x$, we have

$$\langle \phi_{*t}\xi, \phi_{*t}\eta \rangle = \sum_{k=-2}^{2} a_k e^{kt} \tag{4.5.1}$$

for some constants a_k. But if ω is the volume element of $\mathcal{H}^1$ and ξ_i is a basis of $\mathcal{H}^1_x$, then

$$\phi_t^*\omega(\xi_1, \dots, \xi_{2n-k-1}) = (\det\langle \phi_{t*}\xi_i, \phi_{t_*}\xi_j\rangle)^{1/2} \tag{4.5.2}$$

must be constant by (4.5.1) and compactness of $\mathcal{H}^1$.

The flat case is similar: Jacobi fields now have the form $t \mapsto E_1(t) + tE_2(t)$, so that (4.5.2) becomes the square root of a polynomial in t. Compactness of $\mathcal{H}^1$ then forces it to be constant. □

Theorem 4.5.1. *Let M be a compact space form of curvature $\kappa \le 0$. If $\kappa < 0$, then M admits no metric foliations. If $\kappa = 0$, then any such foliation splits; i.e., it is locally congruent to a metric product foliation.*

Proof. We will show that the foliation is flat (and in particular, its orthogonal complement is a totally geodesic foliation). In negative curvature, the statement follows from the fact that compact manifolds of negative curvature admit no totally geodesic foliations [128], and in the flat case, from Theorem 2.2.2.

In the hyperbolic case, consider a point $x \in \mathcal{H}^1$ where the decomposition stated in Oseledets' ergodic theorem holds, so that $\mathcal{H}^1_x = V^s(x) \oplus V^u(x) \oplus V^0(x)$. We claim that

$$\begin{aligned} V^u(x) &\subset \Delta = \{(e,e) \mid e \in M_{\pi(x)}\}, \\ V^s(x) &\subset \Delta^* = \{(e,-e) \mid e \in M_{\pi(x)}\}. \end{aligned} \tag{4.5.3}$$

The arguments are similar in both instances, and we only prove the latter. Denote by $\mathsf{p}^u : \mathcal{H}^1_x \longrightarrow V^u(x)$ the projection. As pointed out earlier, if $\xi = (u,v) \in \mathcal{H}^1_x$, then $Y_\xi(t) = e^t E_1(t) + e^{-t}E_2(t)$, with E_i parallel, $E_1(0) = (u+v)/2$, $E_2(0) = (u-v)/2$. Then

$$\ln|\phi_{t*}\xi| = \ln\left(|Y_\xi|^2 + |Y'_\xi|^2\right)^{1/2}(t) = \ln\sqrt{2} + \frac{1}{2}\ln\left[e^{2t}|E_1|^2 + e^{-2t}|E_2|^2\right].$$

Notice that if $E_1 \neq 0$, then $(\ln|\phi_{*t}\xi|)/t \to 1$ as $t \to \infty$, so that $\mathsf{p}^u\xi \neq 0$. In other words, if $\xi \in V^s(x)$, then $E^1 = 0$, and $v = -u$ as claimed.

Now, consider any horizontal $0 \neq y \perp x$. By Lemma 4.5.1, $(0,y) \in \mathcal{H}^1_x$, and since $V^0(x)$ is spanned by $(0,x)$, $(0,y) = (e,e) + (f,-f) \in V^u(x) \oplus V^s(x)$ for some

e, f. But then $-f = e = y/2$, from which we conclude that $(y, y) \in \mathcal{H}^1_x$. Again by the lemma, $A_y x = 0$. Thus, $A_x = 0$ for almost every x, and by continuity, $\mathcal{F}$ is flat.

In the zero curvature case, choose $x \in \mathcal{H}^1$ as in the statement of Birkhoff's ergodic theorem, with $f : \mathcal{H}^1 \to \mathbb{R}$ given by $f(z) := |A^*_z|^2$. The result will follow once we show that if U is a unit vertical field along the geodesic c in direction x, then $(1/t) \int_0^t |A^*_{\dot{c}} U|^2 \to 0$ as $t \to \infty$. Assume first that U can be written as $J/|J|$ for some holonomy Jacobi field J along c. Since $J = E + tF$ for parallel fields E, F, we have

$$|A^*_{\dot{c}} U|^2 \leq |A^*_{\dot{c}} U|^2 + |S_{\dot{c}} U|^2 = \frac{|J'|^2}{|J|^2} = \frac{|F|^2}{|E|^2 + 2t\langle E, F\rangle + t^2|F|^2},$$

and the claim certainly holds in this case. In general, if J_i, $i = 1, 2$, are holonomy fields with $J_i(0)$ orthonormal eigenvectors of $S_{\dot{c}(0)}$, then the angle $\measuredangle(J_1(t), J_2(t)) \to \measuredangle(J_1'(0), J_2'(0))$ as $t \to \infty$ by linearity of Jacobi fields in Euclidean space. It follows that there exists an orthonormal basis $\{u_i\}$ of eigenvectors of $S_{\dot{c}(0)}$ such that if J_i is the holonomy field with $J_i(0) = u_i$, then the angle between any two $J_i(t)$ and $J_k(t)$ lies in some fixed interval (α, β), for some $0 < \alpha < \beta < \pi$, and all $t > 0$. This in turn implies that U equals a functional linear combination $\sum f_i(J_i/|J_i|)$ with bounded f_i, and thus $(1/t) \int_0^t |A^*_{\dot{c}} U|^2 \to 0$. □

It should be noted that the argument above extends with only minor modifications to compact locally homogeneous manifolds of negative curvature. It is therefore tempting to conjecture that there are no metric foliations on compact manifolds of negative curvature, especially in light of the following result (see [108]):

Theorem 4.5.2. *A compact manifold M with negative Ricci curvature admits no one-dimensional metric foliations.*

Proof. We begin by computing the divergence of the mean curvature vector field $Z = \nabla_T T$ of $\mathcal{F}$, where T is a (local) unit vertical field. Let $p \in M$, $\pi : U \to B$ a submersion defining $\mathcal{F}$ in a neighborhood U of p, and $\bar{X}_i$ local orthonormal fields on B with $\nabla_{\bar{X}_i} \bar{X}_j(\pi(p)) = 0$. Then the basic fields X_i on U that are π-related to $\bar{X}_i$ satisfy $\nabla^{\mathbf{h}}_{X_i} X_j(p) = 0$. Now,

$$\operatorname{div} Z = \sum_i \langle \nabla_{X_i} \nabla_T T, X_i\rangle + \langle \nabla_T \nabla_T T, T\rangle. \tag{4.5.4}$$

The second term on the right equals $-|\nabla_T T|^2$, whereas the first term may be rewritten as

$$\begin{aligned}
\langle \nabla_{X_i} \nabla_T T, X_i\rangle &= X_i \langle \nabla_T T, X_i\rangle - \langle \nabla_T T, \nabla_{X_i} X_i\rangle \\
&= X_i \langle \nabla_T T, X_i\rangle = X_i \langle S_{X_i} T, T\rangle \\
&= \langle \nabla_{X_i}(S_{X_i} T), T\rangle + \langle S_{X_i} T, \nabla_{X_i} T\rangle \\
&= \langle (\nabla^{\mathbf{v}}_{X_i} S)_{X_i} T, T\rangle.
\end{aligned}$$

Using (1.5.9), we obtain

$$\langle \nabla_{X_i} \nabla_T T, X_i \rangle = \langle R((T, X_i)X_i, T\rangle - |A^*_{X_i} T|^2 + |S_{X_i} T|^2.$$

Substituting in (4.5.4) and noticing that $|\nabla_T T|^2 = \sum_i |S_{X_i} T|^2$, we finally get

$$\operatorname{div} Z = \operatorname{Ric}(T) - |A^* T|^2, \tag{4.5.5}$$

with $|A^*T|$ denoting the norm of the operator $x \mapsto A^*_x T$. The theorem now clearly follows, since the divergence of Z integrates to zero over M. □

One further consequence of (4.5.5) is that if the sectional curvature of a compact manifold M is nonpositive, then any one-dimensional metric foliation of M splits: In fact, both the A-tensor and the vertizontal curvatures must vanish, so that Theorem 2.2.2 applies.

In light of the above discussion, a negatively curved manifold M that admits a one-dimensional metric foliation cannot be compact. So what does M look like, topologically? If the curvature is a constant κ, then the answer is known: Namely, when $\kappa = 0$, M must be isometric to $\mathbb{R} \times_\Gamma \mathbb{R}^{n-1}$, where $\Gamma = \pi_1(M)$ acts diagonally by rigid motions. When $\kappa < 0$, M is diffeomorphic to $\mathbb{R} \times (\mathbb{R} \times_\Gamma \mathbb{R}^{n-2})$, with Γ as above, and in particular, M admits a flat metric. For a proof, the reader is referred to [10]. It should be noted, though, that this does not generalize to nonconstant negative curvature. For example, let S denote any compact surface with genus > 1, endowed with a hyperbolic metric, and let N denote the warped product $\mathbb{R} \times_{e^t} S$. Define a function $f : N \to \mathbb{R}$ by $f(t,p) = e^t$. Then the warped product $M = N \times_f \mathbb{R}$ has negative curvature, and since the hypersurfaces $N \times \{t\}$ are totally geodesic in M, their orthogonal complement are the leaves of a one-dimensional metric foliation on M. M, however, is diffeomorphic to $S \times \mathbb{R}^2$.

Bibliography

[1] M. Alexandrino, *Proofs of conjectures about singular Riemannian foliations*, Geom. Dedicata **119** (2006), 219–234.

[2] S. Aloff, N. Wallach, *An infinite family of distinct 7-manifolds admitting positively curved Riemannian structures*, Bull. Amer. Math. Soc. **81** (1975), 93–97.

[3] J. Alvarez López, *The basic component of the mean curvature of Riemannian foliations*, Ann. Global Anal. Geom. **10** (1992), 179–194.

[4] T. Aubin, *Métriques riemanniennes et courbure*, J. Differential Geom. **4** (1970), 383–424.

[5] M. Anderson, *On the topology of complete manifolds of non-negative Ricci curvature*, Topology **29** (1990), 41–55.

[6] M. Anderson, *Short geodesics and gravitational instantons*, J. Differential Geom. **31** (1990), 265–275.

[7] P. Antonelli, D. Burghelea, P. Kahn, *The non-finite homotopy type of some diffeomorphism groups*, Topology **11** (1972), 1–49.

[8] K. Aso, S. Yorozu, *A generalization of Clairaut's theorem and umbilic foliations*, Nikonhai Math. J. **2** (1991), 139–153.

[9] W. Ballman, M. Gromov, V. Schroeder, *Manifolds of Nonpositive Curvature*, Birkhäuser Boston (1985).

[10] A. Basmajian, G. Walschap, *Metric flows in space forms of nonpositive curvature*, Proc. Amer. Math. Soc. **123** (1995), 3177–3181.

[11] V. Bangert, *Totally convex sets in complete Riemannian manifolds*, J. Differential Geom. **16** (1981), 333–345.

[12] Ya.V. Bazaikin, *On one family of* 13*-dimensional closed Riemannian positively curved manifolds*, Sib. Math. J. **37** (1996), 1219–1237.

[13] I. Belegradek, V. Kapovitch, *Topological obstructions to nonnegative curvature*, Math. Ann. **320** (2001), 167–190.

[14] I. Belegradek, V. Kapovitch, *Obstructions to nonnegative curvature and rational homotopy theory*, J. Amer. Math. Soc. **16** (2003), 259–284.

[15] L. Bérard Bergery, *Sur certaines fibrations d'espaces homogènes riemanniens*, Compositio Math. **30** (1975), 43–61.

[16] L. Bérard Bergery, *Les Variétés Riemanniennes homogènes simplement connexes de dimension impaire à courbure strictement positive*, J. Math. Pures Appl. **55** (1976), 47–68.

[17] L. Bérard Bergery, *Certains fibrés à courbure de Ricci positive*, C.R. Acad. Sc. Paris **286** (1978), 929–931.

[18] V. Berestovskii, *Submetries of space forms of nonnegative curvature*, Siberian Math. J. **28** (1987), 552–562.

[19] V. Berestovskii, L. Guijarro, *A metric characterization of Riemannian submersions*, Ann. Global An. Geom. **18** (2000), 577–588.

[20] M. Berger, *Les Variétés Riemanniennes homogènes normales simplement connexes à courbure strictement positive*, Ann. Scuola Norm. Sup. Pisa **15** (1961), 179–246.

[21] J. Berndt, H. Tamaru, *Homogeneous codimension one foliations on noncompact symmetric spaces*, J. Differential Geom. **63** (2003), 1–40.

[22] A. Besse, *Einstein Manifolds*, Springer-Verlag, Berlin Heidelberg, 1987.

[23] R. Blumenthal, J. Hebda, *de Rham decomposition theorems for foliated manifolds*, Ann. Inst. Fourier (Grenoble) **33** (1983), 183–198.

[24] C. Boltner, *On the Structure of Equidistant Foliations of Euclidean Space*, Ph.D. Dissertation, Universität Augsburg (2007).

[25] H. Boualem, *Feuilletages riemanniens singuliers transversalement intégrables*, Compositio Math. **95**, 101–125.

[26] G. Bredon, *Topology and Geometry*, Springer-Verlag, New York, 1993.

[27] T. Bröcker, T. t. Dieck, *Representations of compact Lie groups*, Springer-Verlag New York, 1985.

[28] W. Browder, *Higher torsion in H-spaces*, Trans. Amer. Math. Soc. **108** (1963), 353–375.

[29] D. Burago, Y. Burago, S. Ivanov, *A Course in Metric Geometry*, Graduate Studies in Mathematics **33**, American Mathematical Society (2001).

[30] P. Buser, H. Karcher, *Gromov's almost flat manifolds*, Société Mathématique de France, 1981.

[31] G. Cairns, *Totally Umbilic Riemannian Foliations*, Michigan Math. J. **37** (1990), 145–159.

[32] G. Cairns, R. Escobales, *Note on a theorem of Gromoll-Grove*, Bull. Austral. Math. Soc. **55** (1997), 1–5.

[33] J. Cao, M.-C. Shaw, *The smoothness of Riemannian submersions with nonnegative sectional curvature*, Commun. Contemp. Math. **7** (2005), 137–144.

[34] J. Cheeger, *Some examples of manifolds of nonnegative curvature*, J. Differential Geom. **8** (1972), 623–628.

[35] J. Cheeger, D. Ebin, *Comparison Theorems in Riemannian Geometry*, North-Holland (1975).

[36] J. Cheeger, D. Gromoll, *The splitting theorem for manifolds of non-negative Ricci curvature*, J. Differential Geom. **6** (1971), 119–128.

[37] J. Cheeger, D. Gromoll, *On the structure of complete manifolds of nonnegative curvature*, Ann. of Math. **96** (1972), 413–443.

[38] I.P. Chu, *Riemannian fibrations of Euclidean spaces*, Ph.D. Dissertation, SUNY Stony Brook (1981).

[39] O. Dearricott, J.-H. Eschenburg, *Totally geodesic embeddings of 7-manifolds in positively curved 13-manifolds*, Manuscripta Math. **114** (2004), 447–456.

[40] A. Derdzinski, A. Rigas, *Unflat connections in 3-sphere bundles over* S^4, Trans. A.M.S. **265** (1981), 485–493.

[41] R. De Sapio, G. Walschap, *Diffeomorphism of total spaces and equivalence of bundles*, Topology **39** (2000), 921–929.

[42] J.-H. Eschenburg, *New examples of manifolds with strictly positive curvature*, Invent. Math. **66** (1982), 469–480.

[43] J.-H. Eschenburg, *Freie isometrische aktionen auf kompakten Lie-gruppen mit positiv gekrümmten orbiträumen*, Universität Münster Mathematisches Institut (1984).

[44] J.-H. Eschenburg, *Comparison theorems and hypersurfaces*, Manuscripta Math. **59** (1987), 295–323.

[45] J.-H. Eschenburg, E. Heintze, *Comparison theory for Riccati equations*, Manuscripta Math. **68** (1990), 743–747.

[46] R. Escobales, Jr., *Riemannian submersions with totally geodesic fibers*, J. Differential Geom., **10** (1975), 253–276.

[47] R. Escobales, Jr., *Sufficient conditions for a bundle-like foliation to admit a Riemannian submersion onto its leaf space*, Proc. Amer. Math. Soc. **84** (1982), 280–284.

[48] R. Escobales, Jr., P. Parker, *Geometric Consequences of the Normal Curvature Cohomology Class in Umbilic Foliations*, Indiana U. Math. J. **37**, 389–408.

[49] R. Escobales, Jr., *The mean curvature cohomology class for foliations and the infinitesimal geometry of the leaves*, Diff. Geom. Appl. **2** (1992), 167–178.

[50] M. Falcitelli, S. Ianus, A.M. Pastore, *Riemannian submersions and related topics*, World Scientific, River Edge, NJ (2004).

[51] S. Ferry, *Alexander duality and Hurewicz fibrations*, Trans. Amer. Math. Soc., **327** (1991), 201–219.

[52] E. Ghys, *Feuilletages riemanniens sur les variétés simplement connexes*, Ann. Inst. Fourier (Grenoble) **34** (1984), 203–223.

[53] G. Glaeser, *Fonctions composées différentiables*, Ann. of Math. **77** (1963), 193–209.

[54] A. Gray, *Pseudo-Riemannian almost product manifolds and submersions*, J. Math. Mech. **16** (1967), 715–737.

[55] D. Gromoll, K. Grove, *One-Dimensional Metric Foliations in Constant Curvature Spaces*, in "Differential Geometry and Complex Analysis, H.E. Rauch Memorial Volume", Springer-Verlag (1985), 165–168.

[56] D. Gromoll, K. Grove, *The low-dimensional metric foliations of euclidean spheres*, J. Differential Geom., **28** (1988), 143–156.

[57] D. Gromoll, W. Klingenberg, W. Meyer, *Riemannsche Geometrie im Großen* (Zweite Auflage), Springer-Verlag (1975).

[58] D. Gromoll, W. Meyer, *On complete open manifolds of positive curvature*, Ann. of Math. **90** (1969), 75–90.

[59] D. Gromoll, W. Meyer, *An exotic sphere with nonnegative sectional curvature*, Ann. of Math. **100** (1974), 401–406.

[60] D. Gromoll, W. Meyer, *Examples of complete manifolds with positive Ricci curvature*, J. Differential Geom. **21** (1985), 195–211.

[61] D. Gromoll, K. Tapp, *Nonnegatively curved metrics on $S^2 \times \mathbb{R}^2$*, Geom. Dedicata **99** (2003), 127–136.

[62] D. Gromoll, G. Walschap, *Metric fibrations in Euclidean space*, Asian J. Math. **1** (1997), 716–728.

[63] D. Gromoll, G. Walschap, *The metric fibrations of Euclidean space*, J. Differential Geom. **57** (2001), 233–238.

[64] M. Gromov, *Curvature, diameter, and Betti numbers*, Comm. Math. Helv. **56** (1981), 179–195.

[65] M. Gromov, *Carnot-Carathéodory spaces seen from within*, in Sub-Riemannian Geometry, Birkhäuser Verlag (1996), 79–323.

[66] K. Grove, A. Shiohama, *A generalized sphere theorem*, Ann. of Math. **106** (1977), 201–211.

[67] K. Grove, W. Ziller, *Curvature and symmetry of Milnor spheres*, Ann. of Math. **152** (2000), 331–367.

[68] L. Guijarro, *On the metric structure of open manifolds with nonnegative curvature*, Pacific J. Math. **196** (2000), 429–444.

[69] L. Guijarro, L. Sadun, G. Walschap, *Parallel connections over symmetric spaces*, J. Geom. Anal. **11** (2001) 265–281.

[70] L. Guijarro, P. Petersen, *Rigidity in nonnegative curvature*, Ann. Sci. École Norm. Sup. **30** (1997), 595–603.

[71] L. Guijarro, G. Walschap, *The metric projection onto the soul*, Trans. Amer. Math. Soc. **352** (1999), 55–69.

[72] L. Guijarro, G. Walschap, *Twisting and nonnegatively curved metrics on vector bundles over the round sphere*, J. Differential Geom. **52** (1999), 189–202.

[73] L. Guijarro, G. Walschap, *When is a Riemannian submersion homogeneous?*, Geom. Dedicata **125** (2007), 47–52.

[74] A. Haefliger, *Structures feuilletées et cohomologie à valeur dans un faisceau de groupoïdes*, Comment. Math. Helvet. **32** (1958), 248–329.

[75] A. Haefliger, *Groupoïdes d'holonomie et classifiants*, Transversal structure of foliations (Toulouse, 1982). Astérisque No. 116 (1984), 70–97.

[76] R. Hamilton, *Three manifolds with positive Ricci curvature*, J. Differential Geom. **17** (1982), 255–306.

[77] R. Hermann, *A sufficient condition that a mapping of Riemannian manifolds be a fibre bundle*, Proc. Amer. Math. Soc. **11** (1960), 236–242.

[78] F. Kamber, P. Tondeur, *Foliated bundles and characteristic classes*, Lecture Notes in Math. **493**, Springer Verlag, New York 1975.

[79] F. Kamber, P. Tondeur, *de Rham-Hodge theory for Riemannian foliations*, Math. Ann. **277** (1987), 415–431.

[80] H. Karcher, *Submersions via Projections*, Geom. Dedicata **74** (1999), 249–260.

[81] H. Kim, G. Walschap, *Riemannian foliations on compact hyperbolic manifolds*, Indiana U. Math. J. **41** (1992), 37–42.

[82] S. Kobayashi, *Principal fiber bundles with 1-dimensional toroidal group*, Tôhoku Math. J., **15** (1963), 121–139.

[83] H. Blaine Lawson, *The quantitative theory of foliations*, CBMS Regional Conf. Series **27**, 1977.

[84] J. Levine, *A classification of differentiable knots*, Annals of Math. **82** (1965), 15–50.

[85] G. Meigniez, *Submersions, fibrations, and bundles*, Trans. Amer. Math. Soc. **354** (2002), 3771–3787.

[86] J. Milnor, *On manifolds homeomorphic to the 7-sphere*, Ann. of Math. **64** (1956), 399–405.

[87] J. Milnor, *Curvatures of Left Invariant Metrics on Lie Groups*, Advances in Math. **21** (1976), 293–329.

[88] P. Molino, *Etude des feuilletages transversalement complets et applications*, Ann. scient. Ec. norm. sup **10** (1977), 289–307.

[89] P. Molino, *Feuilletages riemanniens sur les variétés compactes*; champs de Killing transverses, C. R. Ac. Sci. Paris **289** (1979), 421–423.

[90] P. Molino, *Géométrie globale des feuilletages riemanniens*, Nerderl. Akad. Wetensch. A85(1) et Indag. Math. **44** (1982), 45–76.

[91] P. Molino, *Riemannian foliations*, Birkhäuser, Boston 1988.

[92] G.D. Mostow, *Strong rigidity of symmetric spaces*, Ann. Math. Studies **78**, Princeton 1973.

[93] W. Mozgawa, *Feuilletages de Killing*, Collect. Math. **36** (1985), 285–290.

[94] M. Munteanu, *One-dimensional metric foliations on compact Lie groups*, Michigan Math. J. **54** (2006), 25–32.

[95] M. Munteanu, *One-dimensional metric foliations on the Heisenberg group*, Proc. Amer. Math. Soc. **124** (2006), 1791–1802.

[96] J. Nash, *Positive Ricci curvature on fibre bundles*, J. Differential Geom. **14** (1979), 241–254.

[97] B. O'Neill, *The fundamental equations of a submersion*, Michigan Math. J. **13** (1966), 445–457.

[98] B. O'Neill, *Submersions and geodesics*, Duke Math. J. **34** (1967), 363–373.

[99] M. Özaydın, G. Walschap, *Vector bundles with no soul*, Proc. Amer. Math. Soc. **120** (1994), 565–567.

[100] P. Pansu, *Métriques de Carnot-Carathéodory et quasi-isométries des espaces symétriques de rang un*, Ann. Math. **129** (1989), 1–61.

[101] E. Park, K. Richardson, *The basic Laplacian of a Riemannian foliation*, Amer. J. Math. **118** (1996), 1249–1275.

[102] G. Paternain, *Geodesic flows*, Birkhäuser, Boston 1999.

[103] G. Perelman, *Proof of the soul conjecture of Cheeger and Gromoll*, J. Differential Geom. **40** (1994), 209–212.

[104] P. Petersen, *Riemannian Geometry*, Springer-Verlag, New York 1998.

[105] P. Piccinni, *A Weitzenböck formula for the second fundamental form of a Riemannian foliation*, Atti Accad. Naz. Lincei Rend. Cl. Sci. Fis. Mat. Natur. **77** (1984), 102–110.

[106] W.A. Poor, *Some exotic spheres with positive Ricci curvature*, Math. Ann. **216** (1975), 245–252.

[107] A. Ranjan, *Riemannian submersions of spheres with totally geodesic fibers*, Osaka J. Math. **22** (1985), 243–260.

[108] A. Ranjan, *Structural equations and an integral formula for foliated manifolds*, Geom. Dedicata **20** (1986), 85–91.

[109] B. Reinhart, *Foliated manifolds with bundle-like metrics*, Ann. of Math. **69** (1959), 119–132.

[110] B. Reinhart, *Harmonic integrals on foliated manifolds*, Amer. J. Math. **81** (1959), 529–536.

[111] B. Reinhart, *The differential geometry of foliations*, Springer-Verlag, Berlin 1983.

[112] H. Rummler, *Quelques notions simples en géométrie riemannienne et leurs applications aux feuilletages compacts*, Comment. Math. Helv. **54** (1979), 224–239.

[113] J.-P. Sha, D.-G. Yang, *Examples of manifolds of positive Ricci curvature*, J. Differential Geom. **29** (1989), 95–103.

[114] J.-P. Sha, D.-G. Yang, *Positive Ricci curvature on compact simply connected 4-manifolds*, in "Differential Geometry: Riemannian Geometry (UCLA)", PSPM **54** (1990), 529–538.

[115] V.A. Sharafutdinov, *The Pogorelov-Klingenberg theorem for manifolds homeomorphic to* $\mathbb{R}^n$, Sib. Math. Zh. **18** (1977), 915–925.

[116] M. Spivak, *A comprehensive introduction to differential geometry* (Volume 4), Publish or Perish, Boston 1975.

[117] M. Strake, *A splitting theorem for open nonnegatively curved manifolds*, Manuscripta Math. **61** (1988), 315–325.

[118] M. Strake, G. Walschap, *Σ-flat manifolds and Riemannian submersions*, Manuscripta Math. **64** (1989), 213–226.

[119] M. Strake, G. Walschap, *Connection metrics of nonnegative curvature on vector bundles*, Manuscripta Math. **66** (1990), 309–318.

[120] K. Tapp, *Bounded Riemannian submersions*, Indiana U. Math. J. **49** (2000), 637–654.

[121] K. Tapp, *Finiteness theorems for submersions and souls*, Proc. Amer. Math. Soc. **130** (2002), 1809–1817.

[122] K. Tapp, *Conditions for nonnegative curvature on vector bundles and sphere bundles*, Duke Math. J. **116** (2003), 77–101.

[123] K. Tapp, *Flats in Riemannian submersions from Lie groups*, preprint.

[124] Ph. Tondeur, *Foliations on Riemannian manifolds*, Springer-Verlag, New York 1988.

[125] Ph. Tondeur, L. Vanhecke, *Transversally symmetric Riemannian foliations*, Tohoku Math. J. **42** (1990), 307–317.

[126] Ph. Tondeur, *Geometry of foliations*, Birkhäuser Verlag, Basel; Boston 1997.

[127] J. Vilms, *Totally geodesic maps*, J. Differential Geom. **4** (1970), 73–79.

[128] P. Walczak, *Dynamics of the geodesic flow of a foliation*, Ergod. Th. & Dynam. Sys. **8** (1988), 637–650.

[129] N. Wallach, *Compact homogeneous Riemannian manifolds with strictly positive curvature*, Ann. of Math. **96** (1972), 277–295.

[130] P. Walters, *An Introduction to Ergodic Theory*, Springer-Verlag, New York 1982.

[131] G. Walschap, *Nonnegatively curved manifolds with souls of codimension* 2, J. Differential Geom. **27** (1988), 525–537.

[132] G. Walschap, *Metric Foliations and Curvature*, J. Geom. Anal. **2** (1992), 373–381.

[133] G. Walschap, *Foliations of symmetric spaces*, Amer. J. Math. **115** (1993), 1189–1195.

[134] G. Walschap, *Umbilic foliations and curvature*, Illinois J. Math. **41** (1997), 122–128.

[135] G. Walschap, *Geometric vector fields on Lie groups*, Diff. Geom. Appl. **7** 1997, 219–230.

[136] G. Walschap, *Metric Structures in Differential Geometry*, Springer-Verlag, New York 2004.

[137] A. Weinstein, *Fat bundles and symplectic manifolds*, Advances in Math. **37** (1980), 239–250.

[138] B. Wilking, *On fundamental groups of manifolds of nonnegative curvature*, Diff. Geom. Appl. **13** (2000), 129–165.

[139] B. Wilking, *Index parity of closed geodesics and rigidity of Hopf fibrations*, Invent. Math. **144** (2001), 281–295.

[140] B. Wilking, *Positively curved manifolds with symmetry*, Ann. Math. **163** (2006), 1–62.

[141] B. Wilking, *A duality theorem for Riemannian foliations in nonnegative sectional curvature*, Geom. Funct. Anal. **17** (2007), 1297–1320.

[142] H. Yamabe, *On an arcwise connected subgroup of a Lie group*, Osaka Math. J. **2** (1950), 13–14

[143] D. Yang, *On complete metrics of nonnegative curvature on 2-plane bundles*, Pacific J. of Math. **171** (1995), 569–583.

[144] J.-W. Yim, *Distance nonincreasing retraction on a complete open manifold of nonnegative sectional curvature*, Ann. Global Anal. Geom. **6** (1988), 191–206.

[145] J.-W. Yim, *Space of souls in a complete open manifold of nonnegative sectional curvature*, J. Differential Geom. **32** (1990), 429–455.

[146] W. Ziller, *Fatness revisited*, lecture notes (2005).

Index

Progress in Mathematics (PM)

Edited by

Hyman Bass, University of Michigan, USA

Joseph Oesterlé, Institut Henri Poincaré, Université Paris VI, France

Alan Weinstein, University of California, Berkeley, USA

BIRKHÄUSER

Progress in Mathematics is a series of books intended for professional mathematicians and scientists, encompassing all areas of pure mathematics. This distinguished series, which began in 1979, includes research level monographs, polished notes arising from seminars or lecture series, graduate level textbooks, and proceedings of focused and refereed conferences. It is designed as a vehicle for reporting ongoing research as well as expositions of particular subject areas.

PM 272: Barreira, L.
Dimension and Recurrence in Hyperbolic Dynamics (2008). *Winner of the Ferran Sunyer i Balaguer Prize 2008.* ISBN 978-3-7643-8881-2

PM 271: Galicki, K. / Simanca, S. (Eds.)
Riemannian Topology and Geometric Structures on Manifolds (2008, to appear).
ISBN 978-0-8176-4742-1

PM 270: Tschinkel, Y. / Zarhin, Y. (Eds.)
Algebra, Arithmetic, and Geometry. In Honor of Y.I. Manin. Volume II (2008, to appear).
ISBN 978-0-8176-4746-9

PM 269: Tschinkel, Y. / Zarhin, Y. (Eds.)
Algebra, Arithmetic, and Geometry. In Honor of Y.I. Manin. Volume I (2008, to appear).
ISBN 978-0-8176-4744-5

PM 268: Gromoll, D. / Walschap, G.
Metric Foliations and Curvature (2008, to appear). ISBN 978-3-7643-8714-3

PM 267: Barge, J. / Lannes, J.
Suites de Sturm, indice de Maslov et périodicité de Bott (2008). ISBN 978-3-7643-8709-9

PM 266: Pigola, S. / Rigoli, M. / Setti, A.G.
Vanishing and Finiteness Results in Geometric Analysis. In Memory of Alexander Reznikov (2008). ISBN 978-3-7643-8641-2

PM 265: Kapranov, M. / Kolyada, S. / Manin, Yu.I. / Moree, P. / Potyagailo, L. (Eds.)
Geometry and Dynamics of Groups and Spaces (2007). ISBN 978-3-7643-8607-8

PM 264: Miró-Roig, R.M.
Determinantal Ideals (2007)
Winner of the Ferran Sunyer i Balaguer Prize 2007. ISBN 978-3-7643-8534-7

PM 263: Todorcevic, S.
Walks on Ordinals and Their Characteristics (2007). ISBN 978-3-7643-8528-6

PM 262: Fargues, L. / Genestier, A. / Lafforgue, V.
L'isomorphisme entre les tours de Lubin-Tate et de Drinfield. ISBN 978-3-7643-8455-5

PM 261: Stout, E.L.
Polynomial Convexity (2007)
ISBN 978-0-8176-4537-3

PM 260: Holzapfel, R.-P. / Uludağ, A.M. / Yoshida, M. (Eds.)
Arithmetic and Geometry Around Hypergeometric Functions (2007)
ISBN 978-3-7643-8283-4

PM 259: Capogna, L. / Danielli, D. / Pauls, S.D. / Tyson, J.T.
An Introduction to the Heisenberg Group and the Sub-Riemannian Isoperimetric Problem (2007)
ISBN 978-3-7643-8132-5

PM 258: Gan, W.T. / Kudla, S.S. / Tschinkel, Y. (Eds.). Eisenstein Series and Applications (2008). ISBN 978-0-8176-4496-3

PM 257: Lakshmibai, V. / Littelmann, P. / Seshadri, C.S. (Eds.)
Schubert Varieties (due 2009)
ISBN 978-0-8176-4153-5

PM 256: Borthwick, D.
Spectral Theory of Infinite-Volume Hyperbolic Surfaces (2007). ISBN 978-0-8176-4524-3

PM 255: Kobayashi, T. / Schmid, W. / Yang, J.-H. (Eds.)
Representation Theory and Automorphic Forms (2007). ISBN 978-0-8176-4505-2

Progress in Mathematics (PM)

BIRKHÄUSER

Edited by
Hyman Bass, University of Michigan, USA
Joseph Oesterlé, Institut Henri Poincaré, Université Paris VI, France
Alan Weinstein, University of California, Berkeley, USA

Progress in Mathematics is a series of books intended for professional mathematicians and scientists, encompassing all areas of pure mathematics. This distinguished series, which began in 1979, includes research level monographs, polished notes arising from seminars or lecture series, graduate level textbooks, and proceedings of focused and refereed conferences. It is designed as a vehicle for reporting ongoing research as well as expositions of particular subject areas.

PM 254: Ma, X. / Marinescu, G.
Holomorphic Morse Inequalities and Bergman Kernels (2007)
Winner of the Ferran Sunyer i Balaguer Prize 2006.
ISBN 978-3-7643-8096-0

PM 253: Ginzburg, V. (Ed.)
Algebraic Geometry and Number Theory (2006).
ISBN 978-0-8176-4471-7

PM 252: Maeda, Y. / Michor, P. / Ochiai, T. / Yoshioka, A. (Eds.)
From Geometry to Quantum Mechanics (2007).
ISBN 978-0-8176-4512-8

PM 251: Boutet de Monvel, A. / Buchholz, D. / Iagolnitzer, D. / Moschella, U. (Eds.)
Rigorous Quantum Field Theory (2006). ISBN 978-3-7643-7433-4

PM 250: Unterberger, A.
The Fourfold Way in Real Analysis (2006). ISBN 978-3-7643-7544-7

PM 249: Kock, J. / Vainsencher, I.
An Invitation to Quantum Cohomology. Kontsevich's Formula for Rational Plane Curves (2006).
ISBN 978-0-8176-4456-7

PM 248: Bartholdi, L. / Ceccherini-Silberstein, T. / Smirnova-Nagnibeda, T. / Zuk, A.
(Eds.) Infinite Groups: Geometric, Combinatorial and Dynamical Aspects (2006)
ISBN 978-3-7643-7446-4

PM 247: Baues, H.-J.
The Algebra of Secondary Cohomology Operations (2006)
ISBN 978-3-7643-7448-8

PM 246: Dragomir, S. / Tomassini, G., Differential Geometry and Analysis on CR Manifolds (2006).
ISBN 978-0-8176-4388-1

PM 245: Fels, G. / Huckleberry, A.T. / Wolf, J.A.
Cycle Spaces of Flag Domains. A Complex Geometric Viewpoint (2006). ISBN 978-0-8176-4391-1

PM 244: Etingof, P. / Retakh, V. / Singer, I.M. (Eds.)
The Unity of Mathematics (2006)
ISBN 978-0-8176-4076-7

PM 243: Bernstein, J. / Hinich, V. / Melnikov, A. (Eds.)
Studies in Lie Theory (2006)
ISBN 0-8176-4342-3

PM 242: Dufour, J.-P. / Zung, N.T.
Poisson Structures and their Normal Forms (2005)
ISBN 978-3-7643-7334-4

PM 241: Seade, J.
On the Topology of Isolated Singularities in Analytic Spaces (2005). *Winner of the Ferran Sunyer i Balaguer Prize 2005*
ISBN 978-3-7643-7322-1

PM 240: Ambrosetti, A. / Malchiodi, A. Perturbation Methods and Semilinear Elliptic Problems in R^n (2005). *Winner of the Ferran Sunyer i Balaguer Prize 2005.*
ISBN 978-3-7643-7321-4

PM 239: van der Geer, G. / Moonen, B.J.J. / Schoof, R. (Eds.)
Number Fields and Function Fields – Two Parallel Worlds (2005)
ISBN 978-0-8176-4397-3

PM 238: Sabadini, I. / Struppa, D.C. / Walnut, D.F. (Eds.)
Harmonic Analysis, Signal Processing, and Complexity (2005).
ISBN 978-0-8176-4358-4

PM 237: Kulish, P.P. / Manojlovic, N. / Samtleben, H. (Eds.)
Infinite Dimensional Algebras and Quantum Integrable Systems (2005). ISBN 978-3-7643-7215-6

PM 236: Hotta, R. / Takeuchi, K. / Tanisaki, T.
D-Modules, Perverse Sheaves, and Representation Theory (due 2007)
ISBN 978-0-8176-4363-8

PM 235: Bogomolov, F. / Tschinkel, Y. (Eds.)
Geometric Methods in Algebra and Number Theory (2005).
ISBN 978-0-8176-4349-2

PM 234: Kowalski, O. / Musso, E. / Perrone, D. (Eds.)
Complex, Contact and Symmetric Manifolds. (2005)
ISBN 978-0-8176-3850-4

PM 233: David, G.
Singular Sets of Minimizers for the Mumford–Shah Functional (2005)
Winner of the Ferran Sunyer i Balaguer Prize 2004
ISBN 978-3-7643-7182-1

PM 232: Marsden, J.E. / Ratiu, T.S. (Eds.)
The Breadth of Symplectic and Poisson Geometry (2005)
ISBN 978-0-8176-3565-7

PM 231: Brion, M. / Kumar, S.
Frobenius Splitting Methods in Geometry and Representation Theory (2005)
ISBN 978-0-8176-4191-7

MIX
Papier aus verantwortungsvollen Quellen
Paper from responsible sources
FSC® C105338

If you have any concerns about our products,
you can contact us on
ProductSafety@springernature.com

In case Publisher is established outside the EU,
the EU authorized representative is:
Springer Nature Customer Service Center GmbH
Europaplatz 3, 69115 Heidelberg, Germany

Printed by Libri Plureos GmbH
in Hamburg, Germany